The Essentials of Genetics

ADVANCED
BIOLOGY
READERS

The Essentials of Genetics

ADVANCED
BIOLOGY
READERS

R N Jones
A Karp
G Giddings

JOHN
MURRAY

Other titles in this series:

Microorganisms and Biotechnology	0 7195 7509 5
Ecology	0 7195 7510 9

© R. N. Jones, A. Karp and G. Giddings

First published in 2001
by John Murray (Publishers) Ltd, a member of the Hodder Headline Group
338 Euston Road
London, NW1 3BH
Reprinted 2004

Illustrations by Art Construction, Peter Lawrence
Layouts by Eric Drewery
Cover design by John Townson/Creation

Typeset in 11.5/13 pt Goudy by Wearset, Boldon, Tyne and Wear
Printed and bound in Dubai

A catalogue entry for this title is available from the British Library

ISBN 0 7195 8611 9

Contents

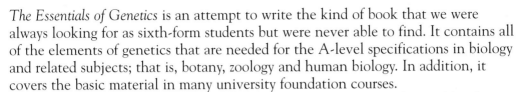

Introduction

The Essentials of Genetics is an attempt to write the kind of book that we were always looking for as sixth-form students but were never able to find. It contains all of the elements of genetics that are needed for the A-level specifications in biology and related subjects; that is, botany, zoology and human biology. In addition, it covers the basic material in many university foundation courses.

The text deals with the factual information required at the Advanced level. It sets out to explain the concepts involved, and gives concise and clear definitions of essential genetics terms. It provides a thematic and integrated approach to the subject. There are worked examples and a wide selection of questions and problems from recent past papers. Further reading is suggested and some links to related websites are provided.

The theme of the book is that genetics is concerned with the *transmission*, the *structure* and the *action* of the material in the nucleus that is responsible for heredity. Wherever possible, we have tried to break the traditional barriers and to integrate the classical parts of the subject with more recent developments at the molecular level. In particular we have drawn together heredity at the level of the chromosome and the DNA molecule. We have also related the molecular basis of gene action and mutation to character differences. In Chapters 16 and 17 we have explained how genes are transmitted at the level of the population and how barriers to free exchange of genes between different breeding groups can lead to the origin of new species. We have not gone into the realms of evolution above the species level.

This book is a revised version of *Introducing Genetics* by R N Jones and A Karp. Chapter 18 on genetic engineering has been extensively revised to take account of some of the changes that have been made in this fast developing field of technology. It now includes a section on the genetic engineering of animals, as well as a discussion of human gene therapy. There is now also an entirely new chapter on the applications of genetic technology to humans – Chapter 19 – including sections on DNA fingerprinting, screening for genetic disorders and the Human Genome Project.

To ease the burden on the reader, the book is organised so that certain more difficult, or esoteric, parts are separated off into boxed sections. We have, though, tried to avoid overloading the student with an excess of information that is either too detailed or simply non-essential. Overall, the material is divided up so that each chapter comprises a natural and self-contained topic or unit, which nevertheless fits within the developing 'story' of the book. The extensive use of cross-referencing between chapters, and the dilution of the text with numerous clear and helpful illustrations, also help to make *The Essentials of Genetics* as accessible and readable as possible.

Acknowledgements

Exam questions have been reproduced with kind permission from the following examination boards:

Assessment and Qualifications Alliance (AQA): The Associated Examining Board (AEB) and the Northern Examinations and Assessment Board (NEAB)
Edexcel Foundation (London Examinations)
OCR (formerly UCLES)

Source acknowledgements

The following is a source from which data has been adapted:

Table 4.3, p.54 R. A. Fisher and F. Yates © 1963, reprinted by permission of Pearson Education Limited

Photo credits

Thanks are due to the following for permission to reproduce copyright photographs:

Cover The Stock Market Photo Agency Inc.; **p.10** Fig. 2.1 *all* R. N. Jones; **p.15** Fig. 2.5a R. N. Jones, Fig. 2.5b Dr David Hayman, Department of Genetics, University of Adelaide; **p.19** Fig. 2.9 *all* R. N. Jones; **p.20** Figs 2.10a, 2.10b & 2.10c Professor John Parker, Botanical Garden, University of Cambridge, Fig. 2.11 Professor Godfrey Hewitt, University of East Anglia; **p.30** Fig 3.1a Biophoto Associates; **p.33** Fig 3.4 Heather Angel; **p.43** Fig 4.1 *left* Roger Scruton, *right* Dr Jeremy Burgess/Science Photo Library; **p.50** Fig. 4.7 Philip Harris Scientific; **p.103** Fig. 9.5 Ed Ricscho/Peter Arnoldine/Science Photo Library; **p.127** Fig. 11.5 Science Photo Library; **p.133** Fig 11.10 *left* Biophoto Associates/Science Photo Library, *right* Megumi Iwano; **p.175** Figs. 14.1a & 14.1b A. Karp; **p.179** Fig. 14.5 *all* Holt Studios, tr Harry Smith Collection; **p.180** Fig. 14.6a & 14.6b Dr Steve Reader, John Innes Institute, Norwich; **p.182** Fig 14.8a Sally & Richard Greenhill, Fig. 14.8b Science Photo Library; **p.203** Fig. 16.1 Science Photo Library; **p.208** Fig. 16.3a Oxford Scientific Films, Fig. 16.3b Charles Numsay & Marie Perennou/Science Photo Library; **p.210** Fig. 16.4a Dr Gopal Murit/Science Photo Library, Fig. 16.4b Eye of Science/Science Photo Library; **p.224** Fig. 17.1a NHPA, Figs. 17.1b, 17.1c & 17.1d A–Z Botanical, Figs. 17.1e, 17.1f & 17.1g Harry Smith Collection; **p.225** Fig. 17.2a NHPA, Fig. 17.2b Animals Unlimited, Fig. 17.2c Oxford Scientific Films; **p.228** Fig. 17.4 NHPA; **p.230** Fig. 17.6 Holt Studios; **p.240** Fig. 18.5 Dr Jeremy Burgess/Science Photo Library; **p.244** Fig. 18.8 Holt Studios.

1   What is genetics?

The word 'genetics' was coined by William Bateson in 1907. He used it to describe a new branch of biology that began in 1900, after the rediscovery of Mendel's work on hybridisation in garden peas. **Genetics** is the science of heredity.

Heredity is the process that brings about the similarity between parents and their offspring; and Mendel had discovered a fundamental law, or rule, about how this process worked (Chapter 3). By 'similarity', we mean that when plants and animals reproduce they have progeny of their own species, and not of some other kind. The offspring of human beings are humans, and not chimpanzees, or rabbits or any other organism.

The members of a family are *similar* to one another in their **characters** (specific characteristics), but *vary* in the details of their individual development and appearance. In the human population of more than four billion people, each of us can be uniquely recognised and distinguished from all the others. These differences between individuals of a family, or of a species, we refer to as **variation**. When we study genetics we want to know how heredity can account for the differences between individuals, as well as for their similarities.

Heredity and environment

It is important to realise that variation has two causes. The differences between individuals are only partly due to the internal factors, inside the cells, that cause heredity. They are also partly accounted for by the external influences of the environment.

In our own species (*Homo sapiens*), it is quite easy to find examples of inherited variation if we look at groups of people from widely separated parts of the world, who have distinctive physical features of the groups to which they belong. The variations include differences in the colour of the skin and hair, texture of the hair, height, shape of the head and facial features. We know that these differences are largely due to heredity because they are known to have been passed on in the same form for several centuries. Moreover, when people migrate and settle in different parts of the world, their descendants retain their racial characteristics regardless of the environments in which they live. On a smaller scale, we can also see the way in which particular characters (colour of hair and eyes, shape of facial features, and so on) run in families within our own ethnic groups.

It is also obvious that not all differences between people are inherited. There are variations that arise due to the level of nutrition, others that are the result of exercise (for example, large muscles) and some that may simply be due to accident or to a whole variety of other environmental causes. There is no doubt that a person who lost a leg in an accident would be able to have children with two legs! Likewise, an Olympic weight-lifting champion, who has built up his muscular physique through exercise and a high protein diet, is no more likely to have muscular offspring than a person who takes no exercise. The development and appearance of a human being – or of a buttercup, a blue tit or any other species – is the outcome of influences due to both heredity and environment.

In some cases it is impossible to say what contribution is made to a particular character by heredity and what is due to environment. This is particularly true where the differences between individuals are small and where a character shows a continuous range of variation within a population (for example, intelligence in humans). In other cases, where the differences between individuals are very clear-cut (for example, male/female, or the presence of an extra finger in humans), we can show that heredity is the cause of the variation by the way in which the character is passed on within a family over successive generations.

Reproduction

Sex cells

The key to understanding genetics lies in knowing what is inherited. What exactly is it that is passed on from the parents to their offspring during sexual reproduction? The answer lies in the **sex cells**, or gametes. Each parent contributes a single gamete to each of its progeny. The male donates the sperm and the female the egg. At fertilisation, the gametes fuse to produce the fertilised egg, or zygote, from which the new individual develops.

Whatever it is that is 'passed on' must be contained in these two tiny cells. They provide the bridge between one generation and the next. A **gamete** is defined as a mature reproductive cell that is capable of fusing with a similar cell of opposite sex to give a zygote. A **zygote** is a cell formed by the fusion of two gametes. The sex cells of humans are shown in Figure 1.1. We will consider them as representative of the gametes produced by all sexually reproducing organisms. The fusion of a sperm cell with an egg cell to form a zygote constitutes the most significant single event in each of our lives.

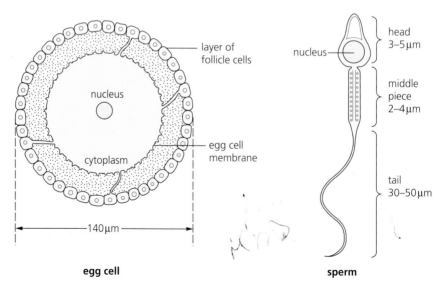

Figure 1.1 Simplified diagram of the gametes of humans

In studying genetics, we will have to look closely at the sperms and eggs. They are the vehicles of heredity. In sexual reproduction it is they alone that carry the genetic programs that determine the similarity of the offspring to their parents, and the heritable differences between one offspring and another. It is during their formation in the male and female reproductive organs that the *processes* that give rise to heredity and variation must take place. In this book we will see what these processes are, and how we find out about them.

Somatic cells

In a multicellular organism, the individual cells that comprise the soma (or body) – that is, the **somatic cells** – all arise by division from the single-celled zygote. These cells are the units of structure and function that comprise all living organisms. The general features of the cell are described in Box 1.1.

Box 1.1 The cell

The cell is the unit of structure and reproduction in all living organisms. There is a wide diversity of types but in all species it conforms to the same basic plan. The substance that makes up the cell as a whole is the **protoplasm**. In plants, this is enclosed within the cell wall.

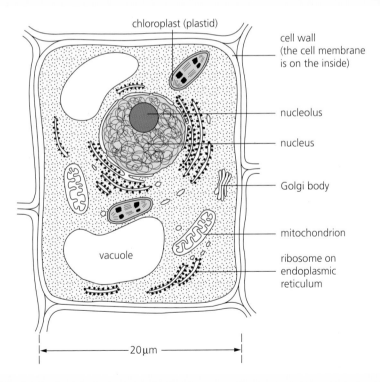

Figure 1.2 The plant cell

The protoplasm is further differentiated into the **nucleus** and the **cytoplasm**. In the cytoplasm are several kinds of organelles and other structures. These can be seen in Figure 1.2, which shows the general features of a plant cell viewed under the electron microscope. Plant cells have two main features that are not present in animal cells – namely plastids and a rigid cellulose cell wall.

The part of the cell in which we are most interested is the nucleus. This is the organelle containing the genetic material responsible for heredity and variation. We will also be concerned, to a much lesser extent, with chloroplasts and mitochondria. These are organelles that have their own genetic information and are able to make more copies of themselves without instructions from the nucleus. Another feature to note is the presence of ribosomes. These numerous small bodies play a key part in the action of the genetic information. They are often associated with the endoplasmic reticulum.

During the multiplication of the body cells, to give a new organism, the processes of heredity must also take place. This has to happen because it is from the somatic cells that the sex organs will be formed, and eventually the next generation of sperms and eggs. In sexual reproduction, the continuity of life proceeds as shown:

$$\text{gametes} \rightarrow \text{zygote} \rightarrow \begin{array}{c}\text{somatic}\\\text{cell}\\\text{division}\end{array} \rightarrow \begin{array}{c}\text{adult}\\\text{organism}\end{array} \rightarrow \begin{array}{c}\text{new}\\\text{generation of}\\\text{gametes}\end{array}$$

All organisms, except some unicellular ones, have to go through this phase of somatic cell division during their development. When the somatic cells divide, their genetic programs are passed on as an exact copy from one cell to another, so that all the cells of the body contain the same complete set of genetic instructions. What makes one body cell different from another one, in its form and function, is the way in which the *different parts* of the genetic program are used in different cells – *not* that some cells have *different programs* to others.

We know that all cells of the body carry a complete set of genetic instructions, because of what happens when some organisms reproduce without sex; that is, **asexually**. This is well known in plants that can propagate vegetatively. A small part of a plant can be detached, as in a geranium cutting, and this can then develop into a new individual. The detached piece clearly contains a complete 'geranium genetic program' in each of its cells. If numerous cuttings are taken from the same parent it will be observed that they are all genetically identical. The only differences that can be seen between them are those due to the environment. Some plants, for example, may receive less light than others and grow more slowly. They will appear different because of the environment, but there are no heritable variations between them.

We shall therefore have to look closely at somatic cell division. We need to know how it differs from the divisions that give rise to the sex cells, and how it provides for heredity without giving rise to any heritable variation.

The nucleus

The fertilised egg, from which a new individual grows and develops, must contain an enormous amount of information in its genetic program. In the case of humans the inherited program of instructions in the cells must be capable of specifying all the details of the structure, the organisation and the functioning of a fully developed person of some 10^{13} cells. Instructions are required to determine the sex of an individual, the details of all the anatomy and physiology of the body, the complexities of the central nervous system and all the characteristics of our form, pigmentation (skin, hair, eye colour) and instinctive behaviour patterns. Each species has its own unique set of instructions, or genetic information within its cells, to determine its own particular pattern of development.

In the study of genetics one of the first questions we want to answer is – *where* is this information located within the cell? Is it in the nucleus or the cytoplasm, or in both? We suspect that it may be in the nucleus. The reason for thinking this is that the sperm and egg make an equal contribution to heredity (as we will see later), yet the sperm contributes far less cytoplasm than the egg (Figure 1.1).

This simple question about the location of genetic information within the cell proved a difficult one to answer. Hämmerling (1943) made the first decisive experiment, using nuclear transplantation, in a single-celled alga called *Acetabularia*. His experiment is described in Figure 1.3.

In later chapters, we see much more evidence to confirm that the genetic information is located mainly in the nucleus. We also discover that some of it is contained in other cell organelles – namely chloroplasts and mitochondria.

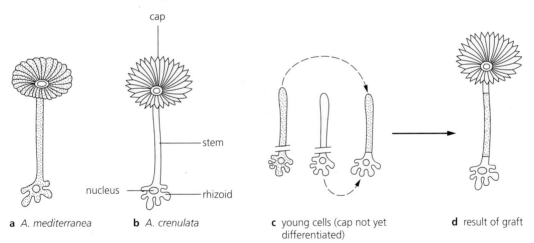

a *A. mediterranea*   b *A. crenulata*   c young cells (cap not yet differentiated)   d result of graft

Figure 1.3 *Acetabularia* is a single-celled alga of tropical seas. It grows to about 6 cm in height and is differentiated into a rhizoid, a stem and a cap. The nucleus is located in the rhizoid. There are two forms – *A. mediterranea* **(a)** and *A. crenulata* **(b)** – which differ in cap morphology. Hämmerling performed a grafting experiment that showed that the genetic information that determined the difference between the two species resided in the nucleus, rather than in the cytoplasm. Young cells were cut in two, and the rhizoid (with nucleus) of the *crenulata* type was grafted onto the stem portion of *mediterranea* **(c)**, and vice versa. The cap that developed was characteristic of the species contributing the nucleus **(d)**

Genetics

Geneticists study the science of heredity, and they are principally concerned with three questions about the information in the nucleus: its *transmission*, its *structure* and its *action*. In addition to these fundamental research aspects, there is also a rapidly growing 'industry' concerned with the applications of genetics to human affairs.

Transmission

How is the genetic information handed on in an exact copy when somatic cells divide? How is it passed on during sexual reproduction so that all the offspring resemble their parents yet differ uniquely from one another? How does its distribution in natural populations change in response to the forces of selection and evolution?

Structure and organisation

What is the chemistry of genetic information? What kind of material carries the information and what is its molecular organisation within the nucleus, the chromosomes and the genes themselves?

Action

How does the genetic material act within the cell to determine the characters, the physiology and the patterns of development of living organisms?

Human affairs

Genetics has always played a large part in human affairs, but not always knowingly so. All of our domesticated plants and animals were selected from wild ancestors, and were much changed in the process, but for several centuries this selection was practised as an 'art' with little or no understanding of its genetic basis. Only after the rediscovery of Mendel's laws of heredity, in 1900, did we begin to systematically alter heredity and to start the scientific breeding of our domesticated plants and animals.

When, in 1953, Watson and Crick discovered the structure of DNA, a whole new area of molecular biology began. The explosion of knowledge that followed pushed genetics into a new dimension. We began to understand what genes are and the mysteries of how they work. In the 1970s the discovery of restriction enzymes – 'molecular scissors' that cut DNA at specific sites – lead to recombinant DNA technology, allowing us to isolate and clone genes, and to manipulate bits and pieces of recombinant DNA. Now our horizons are virtually limitless. We have been able to sequence the whole human genome, and have in our hands the means to clone human genes and even transform ourselves with new genes where our existing ones fail to act properly (gene therapy). We can now clone mammals, and in theory at least see the way ahead to the *unthinkable* experiment of being able to clone human beings. We can modify the genetics of many of our crop plants and domestic animals using recombinant DNA, and transfer genes from one species to another to cause **transgenic organisms** (or cell factories) to make new kinds of proteins, antibiotics or a whole range of pharmaceutical products. Oilseed rape plants can even be made to produce plastic in place of the oil that normally forms as a storage product in their seeds. These exciting new developments bring with them a new burden of responsibility, and the need for moral codes of practice to contain the awesome power that geneticists and biotechnologists now hold.

In this book, we begin by dealing with the transmission of genetic material at the level of the cell and the individual organism (Chapters 2–10). Then we will consider the structure, organisation and action of the genetic material (Chapters 11–14). Chapters 15–16 deal with transmission at the level of the population, and between species during evolution. The last three chapters (17–19) consider aspects of genetics and human affairs; firstly selection imposed by humans, and then the experimental manipulation, utilisation, and engineering of genetic material.

Summary

◆ Genetics is the science of heredity.

◆ Geneticists study the transmission, structure, organisation and action of the material in the cell that is responsible for heredity. These studies are made at all levels of organisation, from molecules and single cells through to individuals and groups of individuals comprising populations.

◆ In the modern world, geneticists can also manipulate and 'engineer' the genetic material in ways that are not found in nature, and the science of heredity is now having a major impact on many aspects of human affairs.

Further reading

1 W. S. Clug and M. R. Cummings (2000) *Concepts of Genetics*. Sixth edition. Prentice Hall, Inc. Website: **http://cw.prenhall.com/bookbind/pubbooks/klug3/**

Mitosis and meiosis

The genetic material responsible for heredity and variation is located mainly in the cell nucleus. One way to study genetics is to observe what happens to the nucleus when a cell divides. There are two kinds of nuclear divisions that we can study – mitosis and meiosis.

Mitosis

Mitosis is division of the nucleus into two daughter nuclei that are genetically identical to one another and to their parent nucleus. It takes place in somatic cells, during development, and is part of the process of cell division. The way that it happens is essentially the same in all **eukaryotes**; that is, in all organisms that have a true nucleus.

To study mitosis, we observe nuclei in cells that are actively dividing. In plants, this means looking at the meristematic regions (growing points) of the roots or shoots. In animals, there are several sources of material. In insects, cells from young embryos can be studied, while in humans and other mammals it is now standard practice to use cell cultures; for example, white blood cells grown in a culture medium.

When we look at a mass of cells, in a piece of squashed-out tissue, we see that they are not usually synchronised in their division, and only a small proportion of them are in mitosis at any one time. The bulk of the nuclei are in **interphase**; that is, the stage in which they are not visibly engaged in division.

The interphase nucleus

The form and appearance of the interphase nucleus can be seen in Figures 2.1 and 2.2 (overleaf). At this stage, it is a spherical structure bounded by a nuclear membrane and containing a mass of chromatin, nuclear sap and one or more spherical nucleoli (see below). The membrane is a double structure and is perforated by a number of pores that allow for the movement of certain materials between the nucleus and the cytoplasm.

Interphase nuclei have a range of sizes. One that is ready to divide contains twice the amount of chromatin, and is about double the size of one that has just completed division. The chromosomes of cells that are about to divide are duplicated during the interphase before division. The copies are then separated, with one of each being incorporated in the two cells resulting from the cell division.

The **chromatin** component is a complex substance made up of approximately equal proportions of deoxyribonucleic acid (**DNA**, which is the genetic material) and protein. Under the light microscope, chromatin has a granular appearance and we are not able to see the details of its structure. The electron microscope shows that it consists of a mass of very thin bumpy fibres with a diameter of about 25 nanometres (2.5×10^{-8} m). The **nuclear sap** is a fluid that is rich in enzymes and various other kinds of molecules concerned with the *activity* of the genetic material.

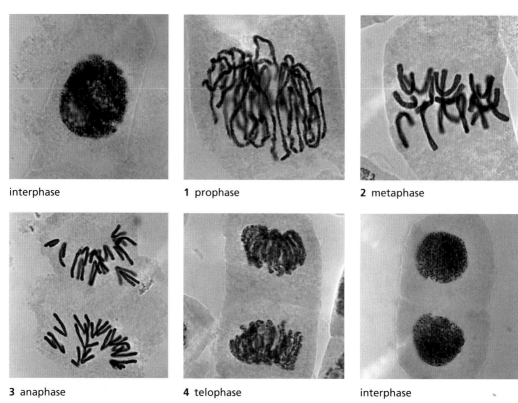

Figure 2.1 Photographs of mitosis in root meristem cells of the onion (*Allium cepa*). *A. cepa* is a diploid species with 8 pairs of chromosomes in its somatic cell nuclei. Interphase nuclei, as they appear just before and just after mitosis, are also shown

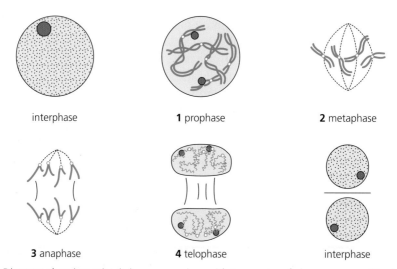

Figure 2.2 Diagram showing mitosis in an organism with two pairs of chromosomes. The interphase nuclei are also shown as they appear just before and just after mitosis. The small spherical bodies in the interphase nuclei, and attached to one pair of chromosomes in prophase and telophase, are nucleoli. Other cell structures are not shown in the diagrams

The **nucleoli** are spherical bodies formed around specialised regions of one (or sometimes more) of the chromosome pairs. These regions are known as nucleolus organisers. Nucleoli have no membrane and consist of an accumulation of proteins and ribosomal ribonucleic acid (**RNA**) molecules that give rise to the ribosomes (Chapter 12).

The interphase nucleus is often misleadingly referred to as being in the 'resting stage' of the division cycle. This is because it shows no visible signs of division at this time, but it does in fact have a high level of metabolic activity (Chapter 12).

Mitotic division

The nucleus undergoes a striking sequence of changes in its form and organisation when it divides during mitosis (Figures 2.1 and 2.2). We know from films of cell division in living material that the sequence is continuous, but for convenience of description we divide the process into a number of named stages, and refer to the form of the chromosomes.

Prophase

In **prophase**, the **chromosomes** make their appearance. The interphase chromatin fibres shorten and thicken, by spiralling and folding, and become visible as chromosomes. Each chromosome is thread-like and consists of two **chromatids** (the sister chromatids) running throughout its length. The chromatids are twisted around one another and joined on either side of an uncoiled region called the **centromere** (or **primary constriction**), as shown in Figure 2.3.

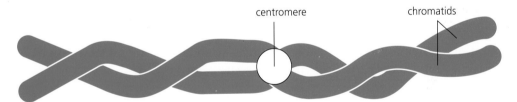

centromere   chromatids

Figure 2.3

Another uncoiled region can be seen at the site of attachment of the nucleolus. This is the **nucleolus organiser** or **secondary constriction** (it is best seen in the metaphase of c-mitosis; see Figure 2.5, page 15). As prophase proceeds the chromosomes become progressively shorter and thicker, and appear as more distinct and separate structures. When they reach their maximum degree of contraction the nuclear membrane breaks down and this marks the end of prophase.

Pro-metaphase

The main event in pro-metaphase is the formation of the **spindle**. As its name implies, this is an ellipsoidal, or barrel-shaped, structure, which is organised in the cytoplasm of the cell. It is broadest at the **equator** and tapers off towards the two extremities, or poles. The spindle is composed of spindle fibres, or **microtubules**.

These are long thin cylindrical filaments assembled from a protein called **tubulin**. Some of the fibres stretch from pole to pole and others run from the poles to the centromeres of the chromosomes. When the spindle is organised, the chromosomes attach to its fibres at random places by their centromeres. The centromeres then move to the equator of the spindle.

Metaphase

At **metaphase** the centromeres are held at the equator of the spindle, under the pull of their spindle fibres, in a position that is equidistant from the two poles. At this stage, as in prophase, the sister chromatids lie closely parallel to one another throughout their length.

Anaphase

During **anaphase**, the two parts of the double centromere (that is, the half-centromeres) begin to separate from one another. The attachment of the sister chromatids, in the regions on either side of the centromeres, then lapses (that is, they cease to be held together) and the chromatids are peeled apart as their centromeres are drawn to opposite poles. Once separated from one another, the sister chromatids are referred to as **daughter chromosomes**.

Telophase

In the final stage of mitosis, **telophase**, the centromeres are drawn together at the poles of the spindle and the nuclear membrane is reformed around the two groups of daughter chromosomes. The prophase coiling sequence is reversed and the daughter chromosomes go back into a diffuse interphase form. Nucleoli also reappear at the sites of their organiser regions. Mitosis is thus complete and we have two daughter nuclei each containing an identical set of single-stranded chromosomes in place of the double-stranded ones that were present in the parent nucleus.

Cytokinesis

After mitosis has ended, **cytokinesis**, or division of the cytoplasm, usually takes place. It is the second part of the cell division process and it separates the cytoplasm, and the newly formed daughter nuclei, into two daughter cells. The mechanism of cytokinesis differs in plant and animal cells. Animal cells undergo *cleavage* by constriction of the cytoplasm and furrowing of the cell membrane. In plants, a cell plate forms across the equator.

Mitotic cycle

The term **mitotic cycle** (or **cell cycle**) refers to the life cycle of an individual cell. It includes all of the events that take place between a given stage of one division cycle and the same stage in the next cycle. We are using the term 'mitotic cycle' here with reference to events going on within the nucleus. The way in which the cycle is divided up, in relation to the division and reproduction of chromosomes, is shown as a diagram in Figure 2.4.

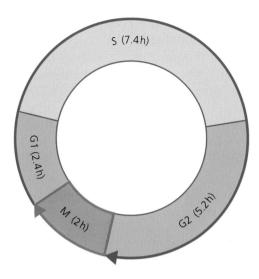

Figure 2.4 Diagram of the mitotic cycle. Times given for the various stages are for meristematic root tip cells of the onion (*A. cepa*), at 20 °C. M = mitosis (prophase → telophase), G1 = gap 1, S = DNA synthesis and chromosome duplication, G2 = gap 2

We know, from various kinds of timing experiments, that mitosis itself occupies only a small part of the total mitotic cycle. The bulk of the period is spent in interphase. In the meristematic root tip cells of the onion, for example, mitosis (M) takes an average of two hours. Interphase occupies 15 h, and this phase is further subdivided into G1 (gap 1), S (period of DNA synthesis) and G2 (gap 2). The chromosomes duplicate themselves, and become double-stranded, during the middle part of interphase.

Details of the molecular events involved in DNA synthesis, and in chromosome structure and duplication, are dealt with in Chapter 11. The duration of the mitotic cycle varies from one species to another. In general its length is related to the amount of genetic material contained within the nucleus. Larger nuclei have longer cycles.

Cell division and cancer

Cancer is a collection of diseases in which something goes wrong with the control of cell division. Cancer cells divide ceaselessly at an inappropriate time, and in an inappropriate place, to form a tumour. If the tumour is **malignant,** cells detach from it and invade other parts of the body to form secondary tumours, a process known as **metastasis**. Some people inherit a **predisposition** for cancer: they are more likely to develop cancer than most other people are. Cancer will of course be more common within the families of such people. Usually, though, cancer is due to an accumulation of **mutations** (damage to the DNA) that can occur at any time of life, but often doesn't result in cancer until a person has become adult. Such mutations may be brought about by exposure to certain chemicals called **carcinogens,** or to too much ultraviolet light (in sunlight).

The cell cycle described in Figure 2.4 is regulated at 'checkpoints' that determine whether the cell will pass from one stage of the cycle to the next. The cycle is normally halted at a checkpoint until critical processes have been completed; for example, DNA synthesis or repair. Genetic defects in this regulation of the cell cycle can cause a cell to become cancerous, or make its descendants more likely to be so. There is, for example, a checkpoint called START, which regulates the transition from G1 to the S phase of the cell cycle. Normal cells will not pass START until any damaged DNA has been repaired. Cells in which the START checkpoint is not working properly will proceed to the S phase with damaged DNA. During a series of cell cycles the replication of damaged DNA will result in more mutations, which may cause further defects in the cell cycle. Cells deriving from one with a dysfunctional START checkpoint can therefore become aggressively cancerous.

C-mitosis

Since chromosomes carry the genetic information, it is of some interest to study their form and to see how they vary in number and size between species. The normal form of mitosis is not ideally suited for this purpose because, in their most contracted state, the chromosomes are tightly grouped around the spindle equator and it is difficult to analyse them individually.

To overcome this inconvenience, we pretreat living cells, before fixing and staining them, with a spindle-inhibiting drug called colchicine. This interferes with the arrangement of protein microtubules (spindle fibres) and destroys the spindle. It also prevents new spindles forming while it is being applied to the cells. As a consequence, mitosis is blocked at metaphase and, because there is no spindle, the chromosomes lie dispersed throughout the cytoplasm. They also become more contracted than in a natural metaphase. This condition is known as **c-mitosis** ('colchicine mitosis') and is ideal for studying metaphase chromosomes. The same effect can be produced by other spindle poisons; for example, alpha-bromonaphthalene.

A c-mitosis from the onion is shown in Figure 2.5. Compare the form of the chromosomes in this photograph with those in the normal metaphase in Figure 2.1.

Chromosomes that have been classified according to their number, form and size are referred to as a **karyotype**. In constructing a karyotype, the positions of centromeres (and of secondary constrictions) can provide a useful marker for chromosome classification. Centromere location is fixed for any one chromosome, but varies for different members of the set. We recognise three main categories of chromosome with respect to their centromere location (Figure 2.6):

1 **metacentric** – centromere approximately in the middle
2 **acrocentric** – centromere near to one end, so that one arm is small and the other much longer
3 **telocentric** – terminal centromere.

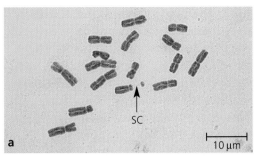
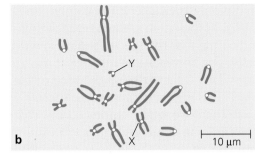

Figure 2.5 Metaphase of c-mitosis in **(a)** the onion (*Allium cepa* var. 'viviparum', micrograph) and **(b)** the red kangaroo (*Macropus rufus*, illustration taken from micrograph). The onion has 8 pairs of chromosomes, which are mostly metacentric (centromeres near the middle). Note the large secondary constriction (SC) in one of the chromosomes. In the red kangaroo, the 10 pairs of chromosomes are mainly acrocentric (centromeres near the end), though there is more variation than in the onion. The cell shown in this illustration is from a male and has an XY sex chromosome pair (Chapter 7)

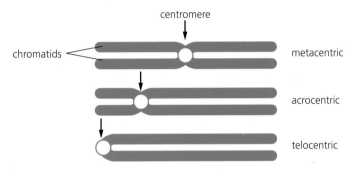

Figure 2.6 The three main forms of metaphase chromosomes classified according to the position of the centromere (arrowed)

Chromosome pairs

In the study of mitosis, and of c-mitosis, we see that the chromosomes are usually present in the cell in identical pairs, which we call **homologues**. The identity of the homologues is most obvious in those species in which pairs have different centromere positions to other pairs within the set (Figure 2.5b). Cells, phases of the life cycle, and organisms that are characterised by having two sets of chromosomes in their cells are said to be **diploid**. The organisms with which we are mainly concerned in this book are all diploid for the major part of the life cycle.

Chromosome numbers

As we have already seen, chromosome numbers vary from species to species. To write down chromosome numbers we use a convention whereby:

x = the **basic number** of different chromosomes in a single (**haploid**) set
n = the **gametic number**, or number of chromosomes in the gametes
$2n$ = the **zygotic number** of chromosomes found in the zygote and in the somatic
    cells derived from it.

15

In diploids, the gametic number (n) and the basic number (x) correspond, so that $n = x$. In the onion (*Allium cepa*), we write the chromosome number as $2n = 2x = 16$, indicating that the species is diploid with two sets of chromosomes in its somatic cells. The chromosome number of the broad bean (*Vicia faba*) we write as $2n = 2x = 12$ and that of the crocus (*Crocus balansae*) as $2n = 2x = 6$. It is necessary to use this system because some species, especially among flowering plants, have more than two sets of chromosomes; that is, they are **polyploid** and have chromosome numbers equal to $3x$, $4x$ and so on, as discussed in Chapter 14. In animals, where diploidy is the general rule, it is usual to give the chromosome number in an abbreviated form; for example, in the red kangaroo (*Macropus rufus*) $2n = 20$, and in humans $2n = 46$ and in the fruit fly (*Drosophila melanogaster*) $2n = 8$.

Mitosis and heredity

What does the study of mitosis tell us about the process of heredity?

1 The first thing we see is that the genetic material in the nucleus is divided up into a number of thread-like bodies that we call the chromosomes.
2 The chromosomes of a set are not all alike. They vary in size and in the positions of their centromeres and secondary constrictions.
3 Chromosomes reproduce themselves in interphase to form two identical sister chromatids.
4 During mitosis, chromatids separate so that one sister chromatid from each chromosome goes into each of the daughter nuclei. Each daughter nucleus therefore contains an identical set of chromosomes.

Heredity in somatic cells is determined by the accuracy of chromosome reproduction and the continuity of the structure of each chromosome from one cell division to the next. Mitosis is a mechanism for distributing identical copies of genetic information to daughter cells during growth and development. In terms of a single chromosome in mitosis the process of heredity can be simply summarised as shown in Figure 2.7.

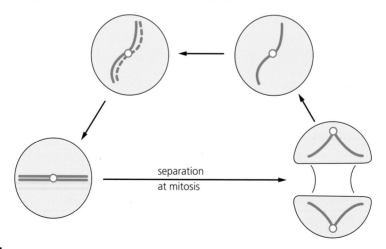

separation
at mitosis

Figure 2.7

Meiosis

Meiosis is the reduction division of the nucleus in which the zygotic number of chromosomes is reduced to the gametic number. This means that in a diploid organism each cell that results from meiosis ends up with only one of each pair of chromosomes, instead of two as in mitosis. In animals, meiosis gives rise directly to the gametes, but in plants it produces spores from which the gametes later arise by mitosis. When two haploid gametes come together at fertilisation, the diploid number of chromosomes is restored again in the zygote. The regular alternation of meiosis and fertilisation in the life cycle therefore maintains the constancy of the chromosome number from one generation to the next.

Meiosis occurs inside the anthers and ovules of flowering plants, and in the testes and ovaries of animals. When studying the process, we usually work with male reproductive organs (anthers or testes) because they have a much greater abundance of cells in meiosis than those of the female.

Visible events of meiosis

The events that take place during meiosis are shown in the sequence of photographs in Figure 2.9 (see page 19), which are from anther squashes in the cereal rye (*Secale cereale*), and in the diagrams in Figure 2.8 (overleaf). As we see, meiosis is more complex than mitosis. It involves two divisions of the nucleus, but the chromosomes only reproduce once. The stages of each of the two divisions are named in the same way as mitosis, except that the first prophase is very much longer and is divided into five sub-stages, as listed in Table 2.1.

Table 2.1

Division I	Division II
prophase I	prophase II
a leptotene	
b zygotene	
c pachytene	
d diplotene	
e diakinesis	
metaphase I (MI)	metaphase II (MII)
anaphase I (AI)	anaphase II (AII)
telophase I	telophase II

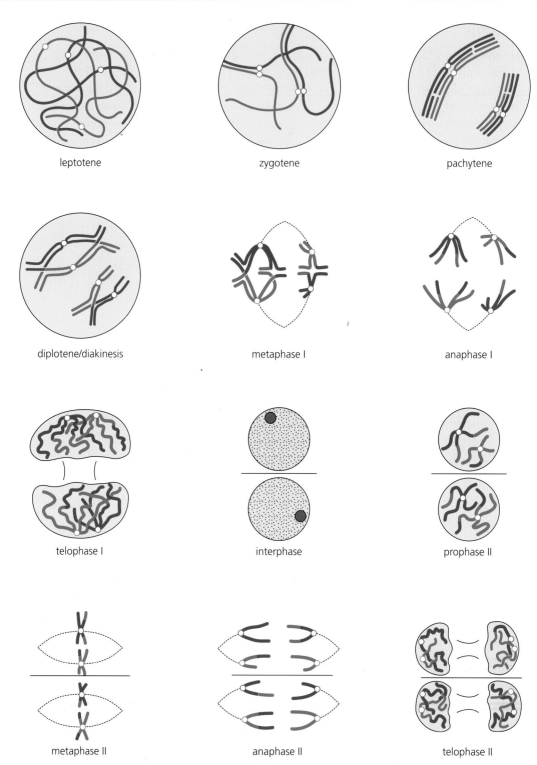

leptotene

zygotene

pachytene

diplotene/diakinesis

metaphase I

anaphase I

telophase I

interphase

prophase II

metaphase II

anaphase II

telophase II

Figure 2.8 Diagram of the stages of meiosis in an organism with two pairs of chromosomes

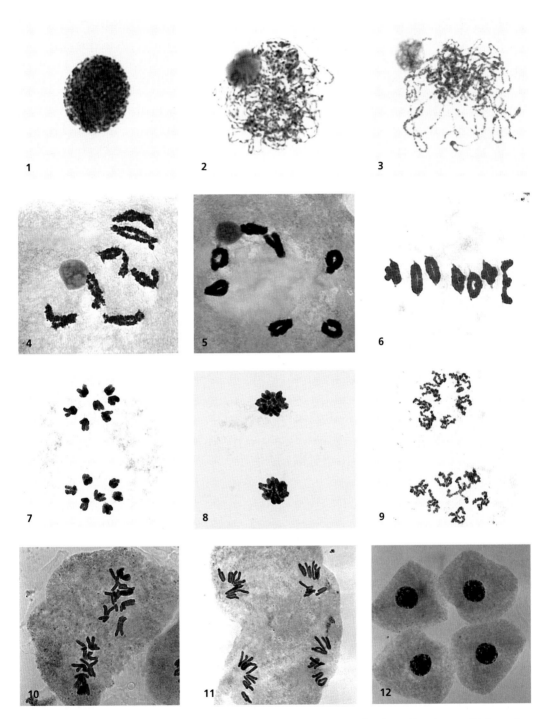

Figure 2.9 Photographs of the stages of meiosis in pollen mother cells of rye (*Secale cereale*, $2n = 2x = 14$). **(1)** Interphase/early leptotene, **(2)** zygotene, **(3)** pachytene, **(4)** diplotene, **(5)** diakinesis, **(6)** metaphase I, **(7)** anaphase I, **(8)** telophase I, **(9)** prophase II, **(10)** metaphase II, **(11)** anaphase II, **(12)** tetrad

Although meiosis proceeds in virtually the same way in all species that have a nucleus, the individual stages can be seen much better in some of them than in others. Details of the early prophase stages (**a, b, c** in Table 2.1) are shown separately in the photographs from the lily in Figure 2.10; those of diplotene are shown in the photograph from the grasshopper in Figure 2.11.

Meiosis is always preceded by an interphase stage during which the genetic material of the nucleus is duplicated in readiness for division. When division begins the chromosomes make their appearance as they shorten and thicken by coiling of their long interphase fibres.

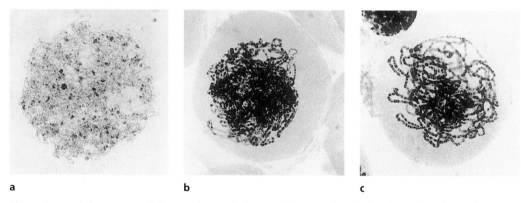

a   b   c

Figure 2.10 Early prophase I stages of meiosis from anther squashes in the flowering plant *Lilium* 'Enchantment': leptotene **(a)**, zygotene **(b)** and pachytene **(c)**

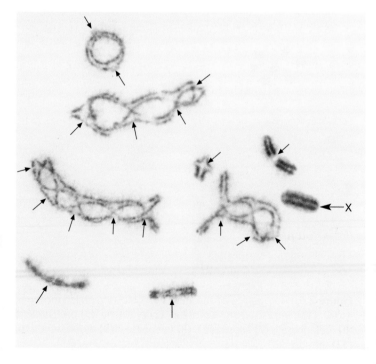

Figure 2.11 Diplotene in the grasshopper *Chorthippus parallelus* ($2n = 16 + X$). The photograph is from a testis squash of a male. The chromatids and the chiasmata (arrowed) can be clearly seen, as well as the single X chromosome. Three of the small bivalents each have a single terminal chiasma (which appears to be in the middle because it is at the ends of the two chromosomes, whose centromeres are at opposite ends)

20

Prophase I

1 Leptotene

The chromosomes become visible as single threads. Although the genetic material has already been duplicated (as we know by measuring its amount), this is *not* evident by light microscopy of the chromosomes. In this respect the early prophase of meiosis differs from that of mitosis.

2 Zygotene

Pairing (or **synapsis**) takes place between homologous chromosomes. It may begin at one or more sites along the chromosomes, and as it proceeds it brings them together in close parallel alignment throughout their length. The pairing is precise and brings corresponding parts of the two homologues into intimate contact.

3 Pachytene

The homologues are now fully paired and contracted into a much shorter and thicker form than in zygotene. **Chromomeres**, bead-like structures due to regions of localised coiling along the chromosomes, are clearly visible at this stage. During pachytene, the chromosomes can be seen to be double-stranded, and form **chiasmata** (singular form is 'chiasma'). Each chiasma is the result of a single break in each of two non-sister chromatids – one from each chromosome – at exactly corresponding places, followed by crosswise rejoining of the broken ends in new combinations (Figure 2.12). This process of breakage and rejoining of chromatids, by which a visible chiasma is formed, is known as **crossing over**. We cannot see these two events happening because of the close and intimate association between the homologues, but the evidence for their occurrence is revealed in the next stage of prophase I.

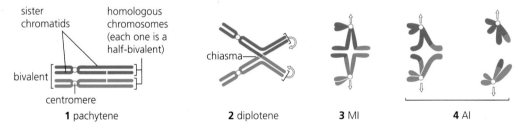

Figure 2.12 A single chiasma is formed, as a result of crossing over. This gives rise to a MI bivalent, which is then resolved when the attraction between the sister chromatids lapses at AI

4 Diplotene

Homologues repel one another and begin to move apart at their centromeres. The force of attraction that held them together until the end of pachytene lapses, but the attraction between the sister chromatids remains. In each **bivalent** (pair of associated homologues) we can now see the chromatids and the chiasmata. This is demonstrated most clearly in the diplotene cell of the grasshopper shown in Figure 2.11 – it is not always so obvious in other species.

The term chiasma refers to the cross-shaped arrangement of the chromatids, which results from crossing over. As far as we can tell, the position of the chiasma corresponds with the point of exchange of the chromatids (see Box 2.1). When a second (or third) chiasma is formed in a bivalent it may involve the same pair, or another pair, of non-sister chromatids, since there are two sister chromatids in each homologous chromosome. It is because of the way in which the sister chromatids attract one another that the two homologues are held together to give the chiasma, following their breakage and rejoining. The chiasma thus serves a mechanical function, which is important in holding homologues together and in providing for their orderly separation in the subsequent stages. It also has a genetic consequence that is explained in the more detailed account of the chiasma given in Box 2.1.

Box 2.1 Chiasmata

Chiasmata are a source of frequent error and confusion. The first thing to understand, in trying to get the better of them, is the way in which they arise by crossing over in prophase I. Breaks occur at precisely corresponding places in the two non-sister chromatids, and then these are rejoined crosswise in new combinations to give a cross-shaped configuration at the point of exchange, as seen in diplotene. In drawing diagrams it is important to remember that the *sister chromatids* always remain in close association, from the time they are formed until their mutual attraction lapses in AI. The diagrams in Figure 2.12 and 2.13 illustrate the formation of chiasmata and the way in which their number and position determine the form of the MI bivalents.

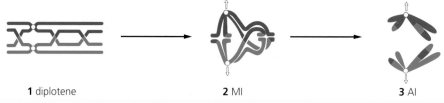

| 1 diplotene | 2 MI | 3 AI |

Figure 2.13 A complex bivalent showing three chiasmata at diplotene and their consequences at MI and AI. (For convenience, all three chiasmata are drawn in the same pair of non-sister chromatids)

It is not normally possible, of course, to see the breakage and rejoining events that are depicted in Figures 2.12 and 2.13, by looking at chromosomes down the microscope. But in organisms where it has been possible to investigate these affairs, it turns out that crossing over does indeed take place at the four-strand stage, after chromosome duplication, and that the points of exchange correspond with the positions of the chiasmata.

The other important point to note about chiasmata is the way in which their formation leads to an exchange of genetic material between the two homologues that pair together to form the bivalents. The consequences of this exchange can be illustrated more clearly in the AI half-bivalents when the chiasmata have been resolved (Figures 2.12 and 2.13). Because the half-bivalents are now composed of two chromatids that are genetically non-identical, the second of the two meiotic divisions is required – in order to separate them into different nuclei. (See *Meiosis and heredity*, opposite.)

Metaphase I

The spindle is organised at pro-metaphase I (the stage before metaphase I) and the bivalents attach to it by their centromeres and move towards the equator. At metaphase I (MI) the homologous centromeres of each bivalent take up positions in which they are equidistant on opposite sides of the equator, and they show signs of being pulled towards the poles.

Anaphase I

At anaphase I (AI) the attraction between sister chromatids lapses, and the half-bivalents are drawn to opposite poles by their centromeres. It is at this stage that the number of chromosomes is reduced from the zygotic to the gametic value; for example, from the diploid to the haploid value in diploid organisms like ourselves.

Telophase I

The chromosomes uncoil and become diffuse again in appearance. In some species they go into an interphase resting stage; in others they do not. Where there is an interphase, it is different from that of mitosis because there is no duplication of the chromosomes. At the end of telophase I the nucleus has completed its first division.

Prophase II

A second cycle of coiling and shortening begins, and the chromosomes are visible again. They differ from those seen in prophase of mitosis in that their chromatids are widely splayed, and do not lie in close parallel alignment. By this stage, of course, the chromatids are not sisters any more – they are genetically non-identical as a result of the crossing over that took place in pachytene.

Metaphase II, anaphase II and telophase II

These stages are similar to those already described for mitosis (see page 12).

At the end of meiosis four haploid nuclei are produced. Each contains a basic set of chromosomes. What happens next depends on the kind of organism concerned, and whether the nuclei have been formed in the male or female reproductive organs.

Meiosis and heredity

Meiosis provides for heredity. It does so by distributing a complete haploid set of chromosomes, and therefore a complete set of genetic information, into each of the gametes. The gametes in turn serve as the vehicles of heredity and transmit their chromosome sets to the zygotes, which are formed at fertilisation. In this way heredity, or the process by which character determinants are handed on from parents to their offspring, takes place.

Meiosis also provides for variation. It does this because the chromosomes distributed to the gametes are not necessarily identical to one another, as they are following mitosis. The products of meiosis are *non-identical*. Precisely what we mean by 'non-identical' will be made much clearer in later chapters. For the moment we simply have to appreciate that meiosis takes place in diploid cells that contain *two sets* of chromosomes – one set originally came from the 'mother' (female sex cell) and one set came from the 'father' (male sex cell) at fertilisation. We refer to these as the 'maternal' and 'paternal' sets. Each set carries the full program of genetic information, but in each set the information that specifies individual character differences (for example, eye colour in humans) may be present in a slightly different form (for example, brown eyes or blue eyes). During prophase I of meiosis, the paternal and maternal chromosomes of each pair come together and exchange parts of their chromatids with one another by crossing over. This occurs in such a precise way that the recombined chromatids neither lose nor gain information, but simply carry new combinations of the different forms of information that were present in the maternal and paternal partners. The non-identical nature of the gametes also depends on how the maternal and paternal chromatids from *different pairs* of homologues are combined together following the reduction division – as explained in Chapter 5.

In dealing with the stages of meiosis, and with its provision for heredity and variation, it is useful to have a simplified scheme that summarises the main features of the process. This is given in Figure 2.14. Table 2.2 summarises the differences between mitosis and meiosis.

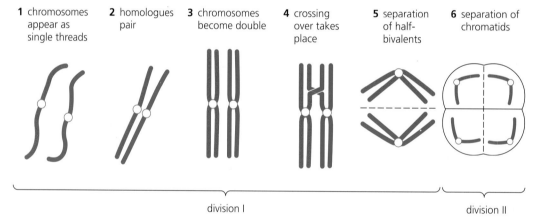

1 chromosomes appear as single threads   2 homologues pair   3 chromosomes become double   4 crossing over takes place   5 separation of half-bivalents   6 separation of chromatids

division I   division II

Figure 2.14 Summary of the main events of meiosis

Table 2.2 Differences between mitosis and meiosis

Mitosis	Meiosis
nucleus divides once, chromosomes once	nucleus divides twice, chromosomes once
two daughter nuclei	four daughter nuclei
chromosome number constant	chromosome number halved
chromosomes double when first seen	chromosomes single when first seen
homologues independent	homologues associate in pairs
products genetically identical	products genetically non-identical

Summary

◆ There are two kinds of division of the nucleus: mitosis and meiosis. Observation of these divisions gives us information about the processes of heredity and variation.

◆ The genetic material in the nucleus is present as a number of thread-like bodies called chromosomes, which maintain their organisation throughout the division cycles.

◆ In the mitotic cycle the chromosomes are duplicated, during interphase, and then their chromatids are separated so that each daughter cell contains an identical set of daughter chromosomes. The cell products of mitosis are therefore genetically identical.

◆ Meiosis is the reduction division that occurs in the sexual cycle of reproduction. It reduces the zygotic ($2n$) number of chromosomes to the gametic (n) number.

◆ Meiosis provides for both heredity and variation. Heredity takes place because each of the cell products contains a complete set of the basic complement of chromosomes. Variation arises from the way in which the cells carry different mixtures of the maternal and paternal chromatids, as well as recombined chromatids that result from crossing over. The products of meiosis are therefore genetically non-identical.

Further reading

1 W. S. Clug and M. R. Cummings (2000) *Concepts of Genetics*. Sixth edition. Prentice Hall, Inc. Website:
  http://cw.prenhall.com/bookbind/pubbooks/klug3/
2 R. N. Jones and G. K. Rickards (1992) *Practical Genetics*. John Wiley & Sons.
3 N. Kleckner (1996) 'Meiosis: How could it work?' *Proc. Natl. Acad. Sci.* **93**: 8167–74.
4 D. Koshland (1994) 'Mitosis: Back to the basics'. *Cell* **77**: 951–4.
5 A. W. Murray and M. Kirschner (1993) *The cell cycle: An introduction*. Oxford University Press.
6 A. Murray and T. Hunt (1993) *The cell cycle*. Oxford University Press.
7 C. J. Sher (1996) 'Cancer cell cycles'. *Science* **274**, 1672–7.

Questions

1 Give an illustrated account of cell division as it occurs in the root and shoot tips of plants.

2 a Describe what happens to a pair of homologous chromosomes during mitosis. Illustrate your answer with diagrams.
  b Describe *two* ways in which mitosis differs from meiosis.

3 a The body cells of horses have 32 pairs of chromosomes; those of donkeys have 31 pairs of chromosomes. How many chromosomes are their in the nuclei of cells of these animals at the end of **(i)** mitosis, **(ii)** the first division of meiosis, and **(iii)** the second division of meiosis?
  b A horse and donkey can be crossed to produce a mule foal. How many chromosomes are there in the body cells of a mule? Explain the reason for your answer.

4 a The chromosome complement of the arctic fox can be described as $2n = 52$; that of the red fox as $2n = 34$. State the number of chromosomes present in the following kinds of cells of each species: **(i)** embryo cells, **(ii)** adult cells, and **(iii)** sperm cells.
  b In captivity, the two species can mate and produce offspring. Describe the chromosome complement of offspring from a cross between the two species.

5 a Explain what a pair of *homologous chromosomes* is.
  b Describe how meiosis leads to genetic variation.
  c State *four* distinct differences between meiosis and mitosis. Briefly discuss the significance of each.

6 a Explain the significance of mitosis in growth.
  b The cell (or mitotic) cycle is made up of interphase and mitosis. Describe the three phases of interphase.
  c Describe the changes in the quantity of DNA that occur in a cell during the cell cycle.

7 a Give an illustrated account of the process of meiosis.
  b Describe how meiosis differs from mitosis and explain the biological significance of each process.

8 a Discuss the biological significance of mitosis. [8]
  b Explain the advantages of using a light microscope, rather than an electron microscope, in studying cells undergoing mitosis. [8]
    OCR (Cambridge Modular) AS/A Sciences: Biology Foundation June 1999 Section B Q2

9 The diagram shows the main stages in the cell cycle.

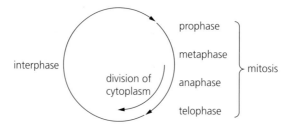

a At what stage in the cell cycle do the following events take place:
   (i) separation of daughter chromatids [1]
   (ii) replication of DNA? [1]
b What evidence would you see on a prepared slide of a bean root tip which would show that metaphase takes longer than anaphase? [1]
c There are 44 chromatids present at the beginning of prophase of mitosis in a rabbit cell.
   (i) How many chromatids would there be in a cell from a rabbit at the beginning of meiosis? [1]
   (ii) How many chromosomes would there be in a sperm cell from a rabbit? [1]

AEB AS/A Biology Module Paper 2 January 1998 Section A Q1

10 The diagrams A–E show stages of mitosis in an animal cell.

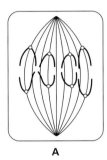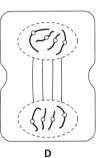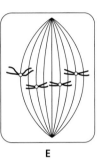

| A | B | C | D | E |

a Which of the drawings A–E shows:
   (i) anaphase
   (ii) telophase
   (iii) metaphase? [3]
b Name *one* stain which could be used to stain the chromosomes in these cells. [1]
c The table shows the average duration of each stage in the cells of a grasshopper embryo.

Stage	Mean duration/minutes
interphase	20
prophase	105
metaphase	13
anaphase	8
telophase	54

Give *one* piece of evidence from the table which indicates that these cells are dividing rapidly. [1]
d Give *two* processes which occur during interphase and which are necessary for nuclear division to take place. [2]

AQA (NEAB) AS/A Biology: Continuity of Life (BY02) Module Test June 1999
Section A Q1

11 Give an account of the process of mitosis. [10]

London AS/A Biology and Human Biology Module Test B/HB1 June 1998 Q8

12 a The table below lists some facts relating to nuclear division. In each case, indicate whether the fact relates to mitosis and/or meiosis by copying the table and placing, in the appropriate column, a tick (✓) if it occurs and a cross (✗) if it does not. The first line of the table has been completed for you. [5]

	Mitosis	Meiosis
takes place at root tips	✓	✗
spindle formed		
homologous chromosomes pair		
chiasmata form		
chromatids separate		
chromosome number of the daughter cells is the same as that of the parent cell		

b Meiosis produces variation in the daughter cells. Explain how
(i) crossing over [3]
(ii) independent assortment [3]
produce variation. (You may use annotated diagrams to assist your explanations.)

OCR (Cambridge Modular) AS/A Sciences: Biology Foundation March 1999 Section A Q2

13 The table shows some statements about mitosis and the two divisions of meiosis. Copy the table, and put a tick in the column(s) where each statement is true. [4]

Statement	Mitosis	Meiosis I	Meiosis II
homologous chromosomes pair together			
centromeres divide and sister chromatids are pulled apart by spindle fibres			
homologous chromosomes are pulled apart by spindle fibres			
occurs in prokaryotic cells			

AQA (AEB) A Biology Paper 1 June 1999 Section A Q7

Mendelian inheritance I: Segregation

Observing chromosomes under the microscope is only one way of studying the transmission of genetic material. Another way is to cross two varieties of a species that differ in some obvious character and study the inheritance pattern of that character from one generation to the next. This kind of breeding experiment was first used successfully by Gregor Mendel in the nineteenth century to unravel the laws of heredity in garden peas. It is a less direct method of experimentation than chromosome cytology because it involves making deductions about abstract 'genetic elements' that control characters, rather than looking directly at the genetic material itself.

In Chapters 3 and 4, we deal with Mendel's breeding experiments and see how they led to the discovery of the laws of heredity. Mendel realised that it is not a character itself that is transmitted during reproduction but some cell element or factor that determines the character. This factor we now call the **gene**.

Design of Mendel's experiments

Mendel set out to discover what happened to two alternative forms of a character when they were combined together to form a **hybrid**. To do this he took two varieties of the garden pea that differed in a single character (for example, 'tall' and 'dwarf' for the character 'length of stem'), crossed them together to make the hybrid, and then observed how the character was passed on over several generations of self-pollination. In this way he found out about the nature of inheritance and the rules by which the transmission of a character could be predicted.

But how was it that Mendel was successful in this objective when others who had tried to do the same before him had failed? It seems that he succeeded because he was an able and careful experimenter. He had a clear understanding of his objectives and of the way in which his experiments should be *designed* in order that he could interpret their outcome.

Choice of organism

The pea, *Pisum sativum*, is an ideal choice of organism for experiments in heredity. There are different types that differ in discrete and easily recognisable characters. The pea is self-pollinating, the anthers and stigma being enclosed within a keel and naturally protected from the pollen of other pea plants (Figure 3.1, overleaf). Controlled crosses are easily made by opening the keel petal, removing the anthers before they shed any pollen, dusting the stigma with pollen from the chosen 'male' parent and then closing the keel for protection. The pea is easily grown and produces an abundance of progeny.

Figure 3.1 (a) Photograph of the flower of the pea, *Pisum sativum*, at the time of pollination

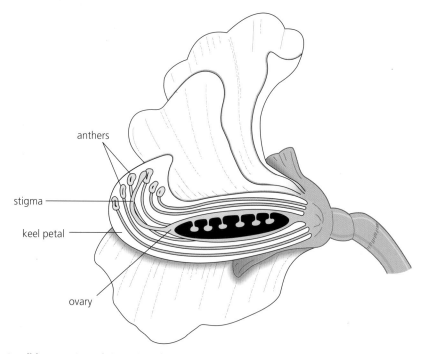

Figure 3.1 (b) Drawing of the side-view of the pea flower. One side of the keel petal is removed to show the way in which the anthers and stigma are enclosed in the keel, which normally results in self-pollination

Determination of ratios

Mendel understood that it was necessary not only to cross together clearly distinct types, but also to count large numbers of offspring over several generations and to establish the *ratios* of their different forms. Earlier workers had made crosses between varieties of peas with contrasting pairs of characters and had noted that a mixture of the two forms appeared in the later generations – but they missed the vital observation, which Mendel made, that these mixtures actually contained the two types in *definite proportions*.

Choice of characters

Mendel confined his studies to seven carefully chosen characters, each of which was represented by discrete alternative pairs (for example, 'tall' and 'dwarf') without any intermediate forms (Figures 3.4, page 33 and 4.1, page 43).

Verification of pure line

Before using any of his pea varieties in crossing experiments, Mendel first established that the seven pairs of alternative characters remained constant when his lines were allowed to self-pollinate naturally for several generations. In other words, he confirmed that for any given character form he had a **pure line**, which was **pure-breeding** (or true-breeding) for that character: seeds from tall plants, for example, always gave rise only to tall plants.

Monohybrid crosses

Mendel crossed together plants that differed in only *one* character; for example, 'round' × 'wrinkled' seeds (Figure 4.1). We now refer to this as a **monohybrid cross**. He then followed the inheritance of round and wrinkled seeds through several generations of self-pollination ('selfing'). In working with this single character – 'seed shape' – he simply ignored the presence of the other six characters. These he tested separately. So, overall, Mendel made seven different monohybrid crosses, for the seven pairs of characters, and studied the inheritance of each character individually. His predecessors had been confused because they had tried to study the inheritance of several pairs of characters all at the same time.

Controls

The other sound measure that Mendel took was to grow uncrossed pure-breeding plants, under the same conditions, to serve as controls when evaluating (or sorting out the characters of) the progeny of the various hybrids.

Monohybrid crosses

The seven characters Mendel chose behaved in the same way in all of the monohybrid crosses, so we may consider just one of them as an example – such as 'length of stem' – and make reference to the other six as appropriate.

The F₁ hybrids

Crosses were made so that each form of the character was used as both 'male' (pollen donor) and 'female' (egg donor) parent. For example, pollen from tall plants was used to fertilise dwarf plants, and vice versa. This is called a **reciprocal cross**:

$$♀ \text{ tall} \times \text{dwarf } ♂$$
$$♂ \text{ tall} \times \text{dwarf } ♀$$

Hybrids from these crosses we now call the 'first filial generation', or F_1. Mendel found that when the seeds were grown up, the F_1 plants were all identical to one another with respect to their height: they all resembled the tall parent and there were no dwarf or intermediate forms. Both the tall and dwarf characters must be present in the hybrid but only the tall character was apparent, or *expressed*. Mendel named this the **dominant** form. The dwarf character, which did not appear, he termed **recessive**. He also found that tall was dominant regardless of whether it had been contributed by the female or the male parent, and that the *direction* of the cross was therefore immaterial (Figure 3.2).

Figure 3.2 (♀ = female; ♂ = male)

The F₂ generation

The 'second filial generation', or F_2, was raised by simply allowing the F_1 hybrids to self-pollinate. In the F_2, the recessive character reappeared, together with the dominant one. Furthermore, when the numbers were counted it turned out that the dominant and recessive forms were in an approximate ratio of 3 dominant : 1 recessive. The same result was found for all seven characters.

The F₃ generation

The breeding behaviour of the F_2 was established by allowing the plants to self-pollinate and then examining a sample of their progeny. Those that showed the recessive character were found to be pure-breeding and to give all dwarf progeny in the 'third filial generation', the F_3. One in three of the tall dominants were also pure-breeding and gave all tall F_3. The other two thirds were classified as 'impure', or 'mixed-breeding' dominants and yielded F_3 families with a 3 : 1 ratio of dominant to recessive types. This breeding pattern, from the parents to the F_3 generation, was found to be the same for all seven characters, and may be summarised in terms of 'tall' and 'dwarf' as shown in Figure 3.3.

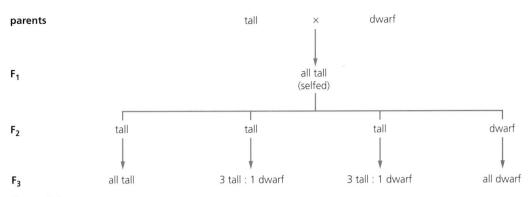

parents   tall   ×   dwarf

F_1 all tall (selfed)

F_2 tall   tall   tall   dwarf

F_3 all tall   3 tall : 1 dwarf   3 tall : 1 dwarf   all dwarf

Figure 3.3

From studying these F_3 progeny, Mendel concluded that the 3 : 1 ratio observed in the F_2 could be further resolved into a 1 : 2 : 1 ratio:

$$
\begin{array}{ccccc}
1 & : & 2 & : & 1 \\
\text{pure-breeding} & & \text{mixed-breeding} & & \text{pure-breeding} \\
\text{dominants} & & \text{dominants} & & \text{recessives}
\end{array}
$$

Figure 3.4 Two-week old pea seedlings showing alternative forms of the character 'length of stem'

Interpretation of the monohybrid crosses

The spark of genius that Mendel showed in interpreting this pattern of inheritance was in seeing that the $1:2:1$ ratio into which the F_2 could be resolved represents what would be expected from randomly combining two pairs of unlike 'elements'.

The Punnett square method, which was later devised by R. C. Punnett, shows us more clearly how this combining process takes place (Figure 3.5).

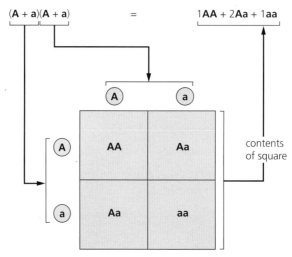

Figure 3.5 Punnett square

The realisation of the way in which the $1:2:1$ ratio in the F_2 could arise gave Mendel the idea that it is not the character itself that is passed on during reproduction, but rather some kind of particulate factor that determines the character and which exists in the cells in two dissimilar forms $(A + a)$. The factors must occur singly in the gametes and are able to combine together randomly in pairs, according to the binomial formula, to give the three F_2 breeding types in the proportions of $1:2:1$. With this idea in mind he was able to interpret the whole of his monohybrid experiment and to arrive at an understanding of the basic mechanism of heredity. The reasoning behind the interpretation, in terms of the cross 'tall' × 'dwarf', is summarised as follows:

1 As both the 'tall' and the 'dwarf' characters reappear in the F_2, there must have been a contribution to the F_1 hybrid by both parents. The F_1 received something, a 'cell element' or 'factor', that determines 'tallness' from the tall parent and a factor that determines 'dwarfness' from the dwarf parent. These cell elements, or factors, we now call genes.
2 Since both forms of the cell element are present in the F_1, the character must be controlled by a pair of elements: one dominant and one recessive. The alternative forms of the cell elements are now called **alleles**.
3 It follows that the cell elements exist in pairs in the cells of the plants, and that in the formation of the gametes only one member of each pair is transmitted into each gamete.

4 In forming the hybrid the two different elements are brought together, without any mixing, or blending, between them, and when the gametes are produced by the F_1 they separate quite cleanly from one another to give two different kinds of gametes. The process by which this separation takes place is now called **segregation**.
5 The F_2 ratio results from random combinations of the two kinds of female and male gametes at fertilisation.

To explain how the F_2 ratios arise, the interpretation is shown in Figure 3.6, overleaf. In this diagram, symbols represent the two forms of the alleles. The bold capital letter (in this case, **T** for 'tallness') denotes the dominant allele and the bold lower-case letter (**t**) represents the recessive allele. The parents are pure-breeding and produce only one kind of gamete, and must therefore contain an identical pair of alleles. We can represent them as **TT** for pure-breeding 'tall' and **tt** for pure-breeding 'dwarf'. Individuals that carry two identical alleles of a gene are **homozygous**. The F_1 hybrids receive one allele from each parent and therefore contain an unlike pair, **Tt** (that is, they are **heterozygous**). They appear tall because **T** is dominant over **t**.

We can see from this method of representation how it is that the $1:2:1$ ratio is the result of the chance, or random, combination of the two kinds of male and female gametes at fertilisation. (Note that the ratio is only likely to be observed when a large enough sample is used. See Box 3.1.) On the male side, we have $\frac{1}{2}$ **T** and $\frac{1}{2}$ **t** in a sample of gametes (that is, half of the gametes produced carry the **T** allele, and half carry the **t** allele). If the half that are **T** have an equal chance of combining with egg cells that are either **T** or **t** from the female side, then half of one half, or one quarter, of the zygotes will be **TT** and half of one half, or one quarter, will be **Tt**. Likewise, the half of the male gametes that are **t** have an equal chance of fusing with **T** or **t** eggs, so that we have a further $\frac{1}{2} \times \frac{1}{2}$ **Tt**, or $\frac{1}{4}$ **Tt**, and $\frac{1}{2} \times \frac{1}{2}$ **tt**, or $\frac{1}{4}$ **tt**.

In observing the F_2, of course, all we see from the appearance of the plants is the ratio of 3 'tall' : 1 'dwarf'. Their actual genetic constitution was determined by allowing them to self-pollinate and by seeing what turned up in a sample of the F_3 progeny. In this way the $1:2:1$ ratio was revealed because the 'dwarfs' yielded all 'dwarf' progeny, confirming that they were homozygous recessive (**tt**); one third of the 'talls' gave all 'tall' F_3 families and must therefore have been pure-breeding homozygous dominant (**TT**); and the other two thirds segregated again to give 3 'tall' : 1 'dwarf' in their F_3 families, thereby confirming their mixed-breeding heterozygous status as **Tt**.

From what has been written above it is obvious that two plants that look identical can be of different genetic constitution, because of the dominance of one allele over another. The genetic constitution of an individual, which is determined by crossing experiments, is called the **genotype**. The appearance of an organism is its **phenotype**. Both **TT** and **Tt** are identical in phenotype ('tall') but they behave differently in genetic crosses, because they are of a different genotype. The environment can also affect the appearance of an organism, as explained in Chapter 1. The phenotype is therefore defined as the appearance and function of an organism as a result of its genotype and its environment.

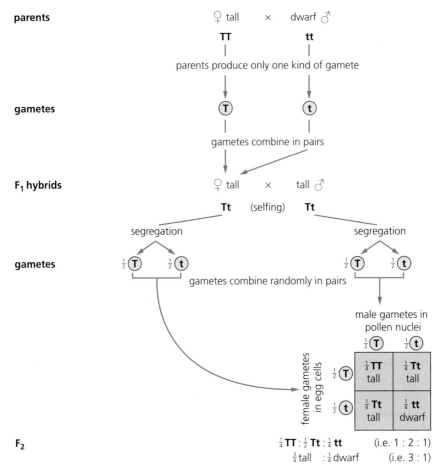

Figure 3.6 Interpretation of Mendel's monohybrid cross between parent pea plants that differ in a single character. In this example, the character is 'length of stem' and the alternative forms are 'tall' and 'dwarf' plants. The pattern of inheritance is explained by postulating the existence of a pair of dissimilar factors in the cells of parents, which are passed on singly through the gametes and combine together in the F_1. One of the factors is dominant over the other, which explains why all the F_1 resemble the 'tall' parent. The factors segregate from one another during sex cell formation in the F_1 to give two kinds of gametes in equal frequencies. Random combination of male and female gametes explains the F_2 ratio of 3 'tall' : 1 'dwarf'. (In the F_1, the cross '♀ tall × tall ♂' refers to male and female parts of the same flower)

The law of segregation

From his interpretation of the monohybrid crosses, Mendel developed his basic theory. He saw that inheritance could only be explained by the existence of character elements in pairs, which are separated in gametes in equal frequencies, and randomly combined back into pairs when the gametes fuse to form the zygotes of the next generation.

Contrary to popular belief, however, Mendel never presented the conclusions of his work in the form of the two laws that are generally attributed to him. It was his successors who redefined his work and made it into the two laws with which we are now familiar. His interpretation of his experiments on the monohybrid crosses thus became known as 'Mendel's first law', or 'the law of segregation'. Because different workers interpreted his laws in their own ways there is no *definitive* version of this law. When we speak of the original form of Mendel's laws, what we are referring to is the redefinition of his work, as others saw it, in terms of his breeding experiments. Put in these original terms, 'Mendel's first law' or 'the law of segregation' may be stated as follows: the characters of an organism are determined by pairs of factors of which only one can be present in each gamete.

Likewise there are numerous modern definitions of this law. These take account of the introduction of new terms and make references to the discovery of the relation between genes and chromosomes, which is discussed in Chapter 5. Stated in modern terms, the **law of segregation** says that: contrasting forms of a character are controlled by pairs of unlike alleles that separate in equal numbers into different gametes as a result of meiosis.

The testcross

To confirm the assumption that '*the various (two) kinds of egg and pollen cells were formed in the hybrids on the average in equal numbers*', Mendel later devised a simple breeding experiment in which his hybrids (**Tt**) were crossed with a pure-breeding recessive homozygote (**tt**). Since the recessive homozygote produces only one kind of gamete, the expectation is that the progeny of the cross will be of two kinds in equal frequencies (Figure 3.7). The actual numbers observed were 87 'tall' (**Tt**) and 79 'dwarf' (**tt**), which Mendel took to confirm the assumption.

This simple breeding test, in which a heterozygote is crossed with the corresponding recessive homozygote in order to verify the genetic constitution of the heterozygote, is now a standard method of genetic analysis and is known as a **testcross**. In the particular case where an individual is crossed with one of its parents, for the same purpose, it is referred to as a **backcross**.

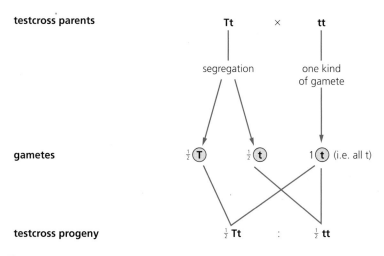

Figure 3.7 The testcross

Box 3.1 Sampling error

In the monohybrid cross **AA** × **aa,** the law or principle of segregation tells us that two kinds of F_1 gametes (that is, gametes from the F_1 generation of **Aa** heterozygotes resulting from the **AA** × **aa** cross) will be produced in equal numbers – **A** and **a**. At fertilisation, they will combine together randomly in pairs to give a ratio of 3:1 of the dominant and recessive phenotypes in the F_2. What we observe in an actual experiment, however, may not always agree exactly with our theoretical expectation. If we have only four F_2 progeny, for example, it is most unlikely that we will find the exact ratio of 3:1. The reason is that *chance* is involved in the production of the two classes of dissimilar gametes from the heterozygotes, as well as in the way in which they come together at fertilisation.

If we were to shake out a small sample of pollen grains from an anther of a plant that is heterozygous (**Aa**), the observed numbers of the two classes of haploid pollen grains, carrying alleles **A** and **a**, may deviate quite widely from the expected 1:1 by chance alone. A sample of only six grains may well contain the unrepresentative numbers of 4 **A** + 2 **a**. Another small sample, by chance, may contain 2 **A** + 4 **a**. In a much larger sample, these small chance deviations will cancel one another out, and the frequencies we find will be much more representative of the actual population of gametes. On the female side, in flowering plants and in animals, only one of the products of meiosis functions as the egg and it is a matter of chance whether it carries the dominant or recessive allele.

At fertilisation, the random combination of male and female gametes to give zygote proportions of 1 **AA**:2 **Aa**:1 **aa** are subject to the same chance effects as we have in tossing two coins and testing for the head/tail combinations HH, HT and TT in the expected ratio of 1:2:1. In other words, our F_2 ratio is subject to **sampling error**, and the smaller the number of progeny we are dealing with the greater the error. For this reason, genetic ratios have to be tested by statistical analysis to see whether the deviation of the observed and expected values is acceptable on the basis of chance, or whether there is some real deviation; that is, the data do *not* fit the expected ratio. These statistical procedures are dealt with in Box 4.2 (see page 53). Mendel was well aware of the problem of sampling error and this is the reason why he worked with such large numbers of F_2 plants.

We can deal with this element of chance by expressing our expectations for a particular cross in terms of probabilities. We say that for the monohybrid cross the probability is 1 in 4 that an individual F_2 will have the recessive phenotype, and 3 in 4 that it will show the dominant phenotype. In this way, we can say what our expectations are even when we have only one offspring. For an *individual*, of course, these expectations are the same regardless of the sample size. The other way of expressing the probability is to say that we expect three quarters of the sample to be of dominant phenotype and one quarter recessive: in this case, though, our chances of fulfilling the expectation will be better when the sample is large.

Mendelian inheritance in other species

Following the rediscovery of Mendel's work at the beginning of the last century, it was quickly found that the basic principles of heredity, which he had discovered in peas, applied to numerous other species of plants and animals as well. In this section we will consider some examples of Mendelian heredity in humans.

Humans

There are a number of familiar and distinctive characters in humans (*Homo sapiens*) that are due to a single gene difference and which show typical Mendelian monohybrid inheritance. Experimental crosses are not possible, of course, but information about heredity can be pooled from different families and from patterns of transmission of characters traced through pedigrees of ancestors and descendants. Once the genetic control of a character has been worked out we can use our knowledge of Mendelian principles to make predictions about the outcomes of certain sexual partnerships and about the probabilities with which particular genotypes will occur among the children.

Albinism is a well-known recessive mutation that is caused by a block in one of the chemical processes leading to production of the pigment melanin. The appearance of homozygotes (phenotype) includes the features of white hair, light-coloured skin and pink eyes. Because the mutation is recessive, it is hidden in heterozygotes (that is, **carriers**) who are capable of transmitting the gene to their progeny. The law of segregation enables us to interpret pedigrees and to make predictions about the phenotypes of children expected from matings between various genotypes. Some examples are described in Figures 3.8 and 3.9.

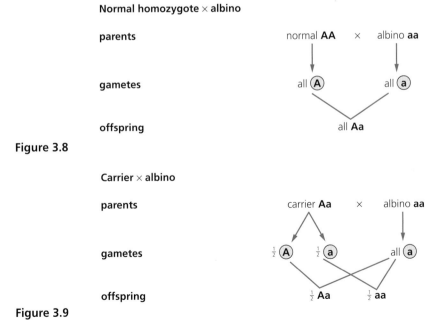

Figure 3.8

Figure 3.9

From the cross shown in Figure 3.8, all the children will be carriers, and will have a normal phenotype. The probability of being normal is 1.0. From the cross in Figure 3.9, half the children are expected to be normal and half albino, but in a small family this expectation may not be realised due to sampling error. A single child has a probability of $\frac{1}{2}$ of being albino. If five albinos are born 'in a row' the sixth one still has a probability of $\frac{1}{2}$ of being normal and $\frac{1}{2}$ of being albino – in the same way that when we toss a coin five times and obtain all heads the probability for the sixth toss is still $\frac{1}{2}$ H and $\frac{1}{2}$ T.

Box 3.2 Tree diagrams

If a couple **Aa** × **aa** (that is, one carrier and one albino parent) intend having only two children, what are the probabilities that:

1 both will be albino
2 both will be normal
3 one will be albino and the other normal?

The answers are worked out as follows. We say that the probabilities for the first child are $\frac{1}{2}$ normal and $\frac{1}{2}$ albino. Now, for each of these possibilities the probabilities for the second child is also $\frac{1}{2}$ normal and $\frac{1}{2}$ albino.

The 'tree diagram' in Figure 3.10 shows that the probability of having two normal children or two albino children is $\frac{1}{4}$. The probability of having one of each is $\frac{1}{2}$, twice as great. Expectations for combinations with more than two children can be worked out in the same way by extending the tree diagram and multiplying through the probabilities.

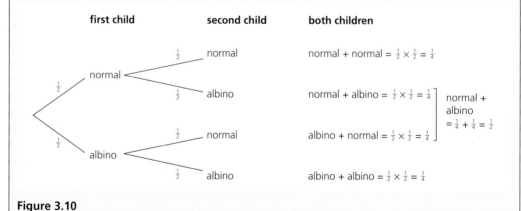

Figure 3.10

If a normal parent of unknown genotype and an albino parent have only two or three children and there is an albino among them, we can be certain that the normal-phenotype parent is heterozygous (that is, a carrier of the recessive **a** allele). If all the children are of normal phenotype, can we be sure the normal parent is homozygous (**AA**)?

Where the pattern of inheritance of a certain character is unknown, its genetic basis can often be worked out for the first time by constructing a pedigree. A **pedigree** is simply a diagram of a family tree over a number of generations showing how the ancestors and descendants are related to one another. The sex of the individuals and their phenotype with respect to a given character can be represented using symbols.

The example shown in Figure 3.11 is a pedigree for brachydactyly (very short fingers) and is one of the first studies of single-gene inheritance in humans. It is an excellent example of the principle of segregation in human genetics.

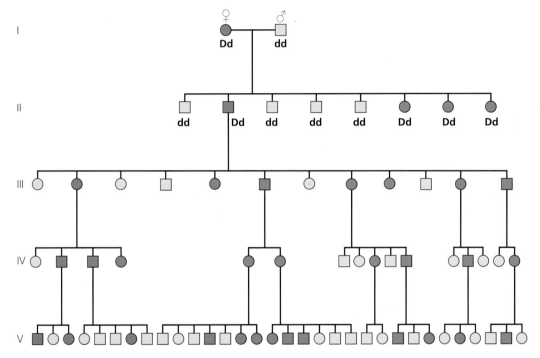

Figure 3.11 Pedigree of the 'short finger' character, brachydactyly, in humans. Females are represented by circles and males by squares. A red symbol indicates an individual displaying the character, while yellow symbols represent normal phenotype. From the pattern of segregation in this family it can be deduced that brachydactyly is a dominant character determined by a single gene. The genotypes are shown for the parents and the next generation only. The spouses, except for the original parents, were all normal (**dd**). They have been left out of the pedigree so that the 1:1 segregation pattern can be seen more clearly

Figure 3.11 shows that the initial partnership between an affected female and a normal male produced eight children in the next generation (labelled II); four were brachydactylous and four were normal. These numbers are reminiscent of a testcross between a heterozygote and a recessive homozygote, and they suggest that the mutation causing brachydactyly is dominant: ♀ **Dd** × **dd** ♂. (If it were recessive, all the children would have been normal.) The suggestion is confirmed by the outcome of the subsequent generations in the pedigree. In each generation, approximately half of the children are brachydactylous and half are normal.

Summary

◆ Mendel discovered that characters are determined by pairs of cell elements, now called genes, which are transmitted from one generation to the next in a regular and predictable manner.

◆ Mendel made his discovery because of the careful way in which he designed his experiments and because his knowledge of mathematics enabled him to interpret his results after counting the numbers of the two kinds of progeny, which segregated out in a fixed $3:1$ ratio in the F_2.

◆ The results of Mendel's monohybrid cross experiments were later formalised as Mendel's first law, or the law of segregation.

Questions follow Chapter 4.

Mendelian inheritance II: Independent segregation

Having resolved the issue of the monohybrid crosses, Mendel wanted to know what happens when *two* pairs of contrasting characters are combined together in the same hybrid. To answer this question, he made crosses between pure lines differing in two characters – that is, **dihybrid crosses**.

Dihybrid crosses

We will consider the experiment in which Mendel followed the inheritance of the two characters 'seed shape' and 'cotyledon colour'. It was already known from the monohybrid crosses that 'round' seed form was dominant over 'wrinkled', and that 'yellow' cotyledon colour was dominant over 'green'. The 'round' and 'wrinkled' phenotypes are shown in Figure 4.1.

Figure 4.1 Mature peas showing alternative forms of the character 'seed shape'

Mendel crossed together parent plants that had round seeds and yellow cotyledons with those having wrinkled seeds and green cotyledons. The F_1 all had the dominant characters of 'round, yellow', and on selfing ($F_1 \times F_1$) they gave rise to an F_2 with four different phenotypes in a ratio $9:3:3:1$. Two of the phenotypes resembled the parents, and two of them were new types combining characters from both of the parents (Figure 4.2).

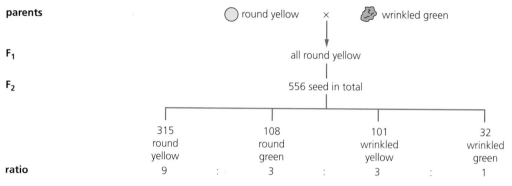

Figure 4.2

In interpreting this result Mendel made the important observation that when the characters of the F_2 phenotypes are looked at individually, the pairs of alternative characters are in the ratio of 3 'round':1 'wrinkled' and 3 'yellow':1 'green' – the same as they were in the monohybrid crosses that he had studied previously. From this observation he saw immediately that the 9:3:3:1 ratio arises simply because the two 3:1 ratios are associated together in the same cross and that the two pairs of characters behave quite independently of one another in their inheritance.

The same theory of combining pairs of cell elements applies as before – but now there are two pairs being transmitted at the same time and this gives rise to four kinds of gametes in both the female and the male, and 4×4 ways in which they can combine together to give the F_2. The interpretation of the experiment is shown in Figure 4.4. The symbols **R** and **r** are used for the dominant and recessive alleles of the gene for seed form, and **Y** and **y** for the dominant and recessive alleles of the gene for cotyledon colour.

The parents are pure-breeding and have the homozygous combinations of alleles **RRYY** for 'round yellow' and **rryy** for 'wrinkled green'. They produce only one kind of gamete: **RRYY** parents produce gametes with **RY** only, and **rryy** parents produce only **ry** gametes. The F_1 are therefore all identical and heterozygous for both pairs of alleles (**RrYy**). Each gamete produced by the F_1 must contain one allele from each gene pair. When the dominant allele **R** passes into a gamete, it has an equal chance of being accompanied by either **Y** or **y**: likewise the recessive **r** also has equal chances of going with **Y** or **y**.

Since **R** and **r** also segregate in equal proportions, four kinds of gametes are produced in equal frequencies (Figure 4.3).

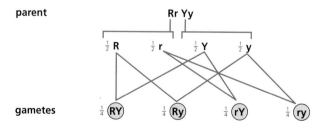

Figure 4.3

The Punnett square method of combining these gametes shows us how the 16 F_2 combinations are formed and how the $9:3:3:1$ ratio of phenotypes arises (Figure 4.4). The essential point to note is that the F_2 ratio depends on there being four kinds of gametes present in equal frequencies, and that the gametes are produced in this way because the two pairs of alleles segregate independently of one another during gamete formation.

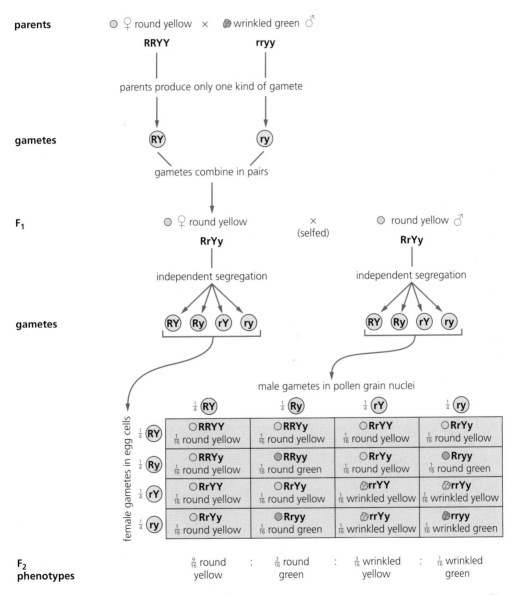

Figure 4.4 Interpretation of the outcome of a dihybrid cross between parent pea plants that differ in two pairs of alleles, which segregate independently of one another from the heterozygous F_1. Random combinations of the four kinds of gametes give rise to the $9:3:3:1$ ratio of phenotypes in the F_2 generation

Confirming the ratio

All that we observe in the F_2 is four phenotypic classes of pea seeds in the ratio of $9:3:3:1$. The existence of the nine different genotypes, and of the four kinds of egg cells and pollen nuclei in equal frequencies, is 'assumed' – as Mendel put it – from the interpretation of the ratio. In order to verify (1) that the nine genotypes were present in their assumed proportions, and (2) that four kinds of male and female gametes were produced in equal frequencies, he carried out two further kinds of breeding tests, in the same way as had been done for the monohybrid crosses:

1 He grew all 556 of the F_2 seeds and then allowed the resulting plants to self-pollinate in order to determine their genotypes by looking at the F_3 progeny. The wrinkled green seeds were found to give plants that were pure-breeding with all wrinkled green seeds, confirming that they were homozygous for both pairs of alleles (**rryy**), as assumed. One third of wrinkled yellow seeds (that is, 28 out of the 96 that grew) gave plants that were pure-breeding for both pairs of alleles, confirming their genotype as **rrYY**, and the other two thirds (68/96) were pure-breeding for 'wrinkled' and mixed-breeding for 'yellow', showing that they were **rrYy**. In the same way, the selfing of 'round, green' and 'round, yellow' gave a mixture of pure-breeding and impure mixed-breeding types, which were in the right proportions to conform with their assumed genotypes.

2 To demonstrate that the different types of gametes were produced by the hybrids in equal frequencies, some F_1 plants were testcrossed to the double-recessive parental type:

F_1 hybrid RrYy × rryy testcross parent

The idea of this testcross is that the double-recessive genotype produces only one type of gamete, carrying the two *recessive* alleles **ry**, and because of this the phenotype of the progeny will *show up* the genotypes of the F_1 gametes from which they were formed. If the assumption about the gametes is correct then the four phenotypes of the testcross progeny should be present in the proportions shown in Figure 4.5. This is exactly what Mendel found. In a sample of 208 progeny there were 55 'round, yellow', 51 'round, green', 49 'wrinkled, yellow' and 53 'wrinkled, green'. These observed values approximate very closely to the expected ones of $52:52:52:52$, for a sample of that size, and give direct confirmation that the four kinds of gametes are produced in their assumed frequencies.

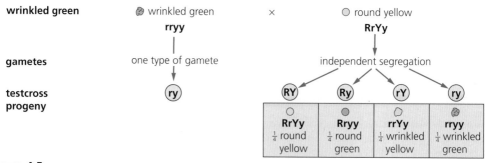

Figure 4.5

The law of independent segregation

Mendel continued his studies by making experiments involving three pairs of contrasting characters. He found that the characters behaved in the same independent way and that all combined randomly with one another in the F_2. In general terms, three pairs of alleles in a cross, **AABBCC** × **aabbcc**, give a trihybrid F_1, which produces eight different gametes, and on selfing gives an F_2 with eight phenotypes in the ratio of $27:9:9:9:3:3:3:1$.

Box 4.1 Sampling error and probability

The question of sampling error arises in dihybrid crosses, just as it did in monohybrid crosses discussed in Box 3.1 (see page 38).

In a very small sample of F_2 individuals, it is meaningless to say that we expect to get the four phenotypic classes of progeny in the exact proportions of $\frac{9}{16}:\frac{3}{16}:\frac{3}{16}:\frac{1}{16}$. We may have fewer than 16 progeny – then what are our expectations? The answer is to use probabilities. We simply have to say for an individual seed the probability that it will be 'green, wrinkled' is 1 in 16, that it will be 'yellow, wrinkled' is 3 in 16, and so on. The probabilities can be obtained from the Punnett square (Figure 4.4), or by simple multiplication of the probabilities for the separate characters. The expectation:

for 'round' and 'wrinkled' is $\frac{3}{4}$ round : $\frac{1}{4}$ wrinkled
for 'yellow' and 'green' is $\frac{1}{4}$ yellow : $\frac{1}{4}$ green

As the two characters are independent of one another, the probabilities for the various combinations of them are the product of their separate probabilities:

'round, yellow' $= \frac{3}{4} \times \frac{3}{4} = \frac{9}{16}$
'round, green' $= \frac{3}{4} \times \frac{1}{4} = \frac{3}{16}$
'yellow, wrinkled' $= \frac{3}{4} \times \frac{1}{4} = \frac{3}{16}$
'green, wrinkled' $= \frac{1}{4} \times \frac{1}{4} = \frac{1}{16}$

What is the probability of getting pure-breeding 'round, green' (**RRyy**)?

The probability of finding the **RR** genotype for the seed form character is 1 in 4, because in the F_2 we have:

$$\frac{1}{4} \textbf{RR} + \frac{1}{2} \textbf{Rr} + \frac{1}{4} \textbf{rr}$$

The probability of getting pure-breeding 'green' (**yy**) is also 1 in 4:

$$\frac{1}{4} \textbf{YY} + \frac{1}{2} \textbf{Yy} + \frac{1}{4} \textbf{yy}$$

The probability of obtaining pure-breeding 'round, green' is therefore:

$$\frac{1}{4} \textbf{RR} \times \frac{1}{4} \textbf{yy} = \frac{1}{16} \textbf{RRyy}$$

which is 1 in 16 (as in the Punnett square).

The experiments on the dihybrid and trihybrid crosses confirmed Mendel's theory. He saw that the way in which the unlike pairs of factors in a hybrid could separate from one another into the gametes and then come together again in predictable combinations to give the F_2 was the same regardless of the number of different character pairs involved in the cross.

This part of his work later became known as 'Mendel's second law' – 'the law of independent assortment' or 'the law of independent segregation'. There is no definite version of this law, but defined in terms of the breeding experiment results alone it can be stated as follows: when two or more pairs of characters are brought together in a cross they segregate independently of each other. Expressed in present day terminology, the 'law' is somewhat different and now takes account of the known chromosomal basis of heredity. The **law of independent segregation** is the most widely used term and it may be defined thus: two or more unlike pairs of alleles segregate independently of each other as a result of meiosis, provided the genes concerned are unlinked. In Chapter 5 we explain the chromosomal basis of independent segregation and how it is that the law only applies to genes that are *unlinked*; that is, located on different pairs of chromosomes.

Recombination

In Mendel's dihybrid cross of 'round, yellow' × 'wrinkled, green' we have already mentioned that two of the F_2 phenotypes resembled the original parents ('round, yellow' and 'wrinkled, green') and that two of them displayed new combinations of characters not shown by either parent ('round, green' and 'wrinkled, yellow'). The process by which new combinations of parental characters may arise is known as **recombination**, and the individuals that possess the new combinations are called **recombinants**. Recombination is important because it is one of the factors that leads to genetic variation in natural populations – about which we will have more to say in Chapter 6.

The term 'recombination' refers to the phenotype of an individual and recombinants are always classified with respect to the parental types originally used in the cross. In the cross 'round, yellow' × 'wrinkled, green' for example, the two pairs of characters are combined into the F_1, which resembles the 'round, yellow' parent (because of dominance), and then recombined during gamete formation in the F_1 to give the two new phenotypes in the F_2 (Figure 4.6).

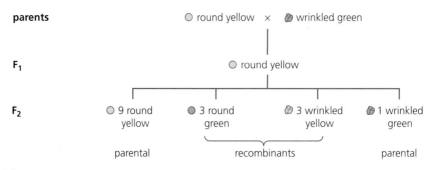

Figure 4.6

Importance of Mendel's work

The importance of Mendel's work received recognition from his successors who redefined it in their own terms and gave us what are now known as 'Mendel's laws of heredity'. The 'laws' are also referred to as 'the first and second principles of genetics'.

Why is Mendel's work so important?

Mendel's work is of enormous importance, firstly because he gave us the method of genetic experimentation and explained how heredity works.

Segregation simply means that a heterozygote, **Aa**, produces two kinds of gametes in equal frequencies, of $\frac{1}{2}$ **A** and $\frac{1}{2}$ **a**. Pure-breeding homozygotes do not segregate; they only produce one kind of gamete: **AA** gives all **A**, and **aa** gives all **a**. When two or more genes are involved in a cross they segregate independently. If an individual is heterozygous for both gene pairs, **AaBb**, we get four classes of gametes in equal frequencies – $\frac{1}{4}$ **AB**, $\frac{1}{4}$ **Ab**, $\frac{1}{4}$ **aB**, $\frac{1}{4}$ **ab**. When only one of the two pairs is heterozygous then only that pair segregates and we get two classes of gametes in equal frequencies – **AaBB** gives $\frac{1}{2}$ **AB** and $\frac{1}{2}$ **aB**. Using these basic rules for one or more pairs of genes, it is a simple matter to obtain the kinds and frequencies of gametes for whatever pair of individuals are involved in the cross, and then to find their random combinations by the Punnett square method. (Remember, though, that the rules only apply to unlinked genes on different chromosomes.)

The second reason why Mendel's work is so significant is that it meant people could now understand the nature of heredity, whereas previously they had no proper idea about it. Before 1900, there was a widely held view that gametes contained 'essences', which were derived from cells of various parts of the body, and that when fertilisation occurred they were blended together in some way to give the form of the progeny. **Blending inheritance** was held to explain how offspring could show some of the characters of both of their parents, but it was inconsistent with the observation that characters could turn up in later generations of a cross in an 'undiluted' form. Mendel's work gave rise to a completely novel idea, and to what is probably the most important concept ever formulated in genetics – that of **particulate inheritance**. According to this idea, it is not the character itself that is transmitted during reproduction, but some discrete particles within the cells that determine each of the characters. These particles maintain their identity and 'separateness' during inheritance through a hybrid and are handed on from one generation to the next in an unmodified and constant form. In anticipating the existence of such particles, Mendel discovered the units of heredity that Johannsen later (1909) christened 'genes'. Mendel's method of experimentation also laid the foundation of what Bateson was later to call the science of 'genetics'.

Introducing *Drosophila*

The fruit fly, *Drosophila melanogaster* (Figure 4.7), is one of the most intensively studied of all 'genetic organisms'. It completes its life cycle in the short duration of only 10 days at 25 °C, it is easy to handle in the laboratory and has many excellent characters that can be used to identify its genes. It has four pairs of chromosomes ($2n = 8$).

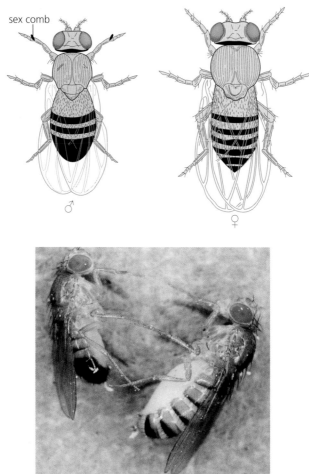

Figure 4.7 Male and female specimens of the fruit fly, *Drosophila melanogaster*. The female is larger than the male and has a lighter coloured abdomen. A diagnostic feature of the male is the sex comb (tuft of bristles) on the front leg

Drosophila geneticists use internationally agreed names for the mutant forms of various characters. A **mutant** is an inherited departure from the normal phenotype. Each mutant character is given a name describing its characters – this is usually an adjective such as 'ebony' for ebony-coloured body. All of the *Drosophila* mutants we shall be referring to in this book are recessive (except for Bar, which is dominant). A description of some well-known mutant characters in *Drosophila* is given in Table 4.1.

Table 4.1 Examples of mutant characters in *Drosophila*

Name	Description
white	white eye colour
miniature	small wings
brown	brown eye colour
vestigial	vestigial wings
ebony	ebony body colour
bar	eye reduced to a narrow slit

Example of a dihybrid cross in *Drosophila*

A pure-breeding female fruit fly with the normal phenotype of normal wings and grey body was mated with a pure-breeding male fly with vestigial wings and ebony body. The F$_1$ were all normal in appearance and when mated together gave an F$_2$ with four classes in the proportions shown in Figure 4.8.

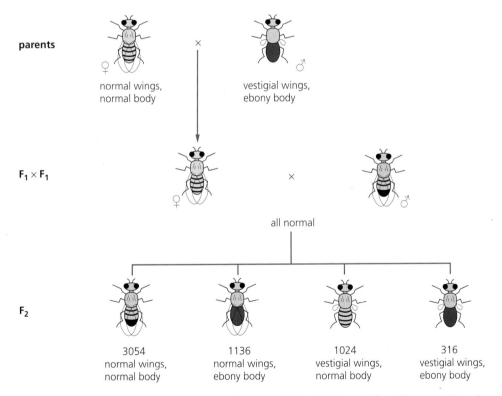

Figure 4.8 The result of an experiment in which a normal female *Drosophila*, with normal length wings and grey body, was crossed with a mutant male with vestigial wings and ebony body. What can we deduce about the inheritance of the two characters 'wing form' and 'body colour'?

How can we explain the results in Figure 4.8 in terms of the inheritance of the characters 'wing form' and 'body colour'?

The F$_1$ has the normal phenotype and resembles the female parent. We deduce that the characters 'vestigial' and 'ebony' are recessive. In the F$_2$, there are four phenotypic classes, which are suggestive of a 9:3:3:1 ratio – an indication that this is so can be derived by dividing each class by the lowest one, which is 316. Expected numbers for a 9:3:3:1 from a sample of 5530 F$_2$ flies are 3111:1037:1037:345, which fit the observed numbers very closely.

The interpretation that we make, from our knowledge of Mendel's second law, is that each of the characters is determined by a single gene with two alleles and that in the gametes of the heterozygous F$_1$ the two pairs of alleles are segregating independently of one another (Figure 4.9).

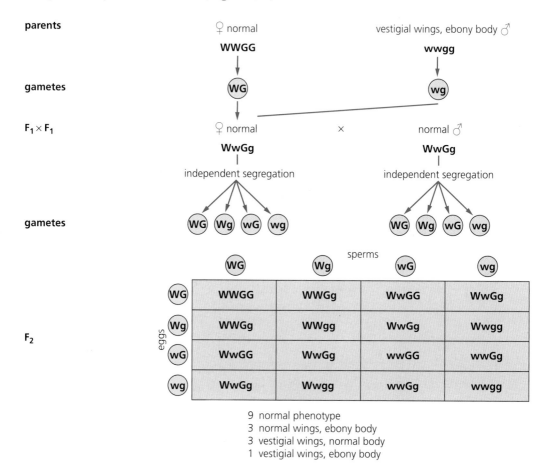

9 normal phenotype
3 normal wings, ebony body
3 vestigial wings, normal body
1 vestigial wings, ebony body

Figure 4.9

In this example it is quite obvious that the observed numbers of the four classes of F$_2$ phenotypes agree closely with those that we would expect for a 9:3:3:1 ratio. But this may not always be the case. We may be uncertain in some situations as to whether an observed set of experimental results do, or do not, fit a particular genetic ratio. To help us make a decision, we use the statistical test outlined in Box 4.2.

Box 4.2 Chi-squared test

As explained in Boxes 3.1 and 4.1 (pages 38 and 47), chance events are involved in the processes of heredity, and experimental results are subject to sampling error. This means that the numbers of individuals we actually observe in the various phenotypic classes of our F_2 or testcross progenies will deviate to some extent from those that we expect to find. If the deviations between the observed and expected values are large there may be some difficulty in deciding whether the results obtained agree with the ratio concerned or not. Mendel clearly recognised this problem and overcame it by working with large samples of F_2 plants – more than 8000 in one case. He had to do this because there were no statistical methods available at the time for assessing the reliability of his data.

The problem is straightforward. Consider the progeny from a cross between two red-flowered pea plants, **Rr** × **Rr**, where **R** is the dominant allele that determines red flowers and **rr** gives white. We expect the progeny to show a 3:1 ratio of red:white flowered plants. If we observe 152 red and 48 white, we will be quite satisfied with the result (the expected values being 150 and 50 for a sample of 200). But suppose we find (1) 156 red and 44 white, or (2) 165 red and 35 white. Is either, or both, of these outcomes acceptable as a 3:1 ratio?

The answer is that 156 red and 44 white *is* in agreement with our hypothesis of a 3:1 ratio, but 165 and 35 is *not*. We arrive at this conclusion by the use of a simple statistical procedure known as the **chi-squared test** (χ^2 test). Essentially, the χ^2 value that we calculate is a measure of the size of the difference (deviation) between the *observed* and *expected* results. The test takes account of the size of the sample, as well as the deviation, and gives the answer as a single numerical value. We use this value to assess how likely it is that the deviations we observe can be accounted for by *chance*, or whether they represent a *real departure* from the expected ratio.

The formula for calculating χ^2 is:

$$\chi^2 = \Sigma \frac{(O - E)^2}{E}$$

where Σ = sum of, O = observed values, and E = expected values.

In the example (1) given above, the observed values are 156 red and 44 white. The total size of the sample is 200 plants. The expected values for a 3:1 ratio are therefore 150 and 50 (that is, $\frac{3}{4} \times 200$ and $\frac{1}{4} \times 200$). The deviation (difference) of the observed values from those expected are $156 - 150 = 6$ for red, and $44 - 50 = -6$ for white. In each case the square of the deviation is 36, and this square divided by the expected numbers gives $36/150 = 0.24$ for red and $36/50 = 0.72$ for white. The χ^2 value is found by adding these two values together to give 0.96, as shown in Table 4.2.

Table 4.2

	O	*E*	Deviation (*d*) *O – E*	*d²*	*d²/E*
red	156	150	+6	36	0.24
white	44	50	−6	36	0.72
	200	200	0		$\chi^2 = 0.96$

We now consult a table of the distribution of χ^2 (Table 4.3) to find out what the probability (P) is of obtaining a deviation as large as, or larger than, the one we have by *chance alone*. The table takes account of the fact that the size of the χ^2 depends upon the number of independent comparisons (phenotypic classes) that contribute to its value, as well as the size of the deviations in the individual classes (in this case two classes, red and white). If we are summing over four classes, as in the third example (below), then the χ^2 will be larger simply because we are adding four numbers rather than two.

We allow for the number of independent comparisons involved in our test by using 'degrees of freedom' (d.f.), which appear down the left-hand side of the table. Two classes (red and white) have only one degree of freedom because there is only one independent comparison; that is, given the number in one class the number in the other class is fixed for a constant sample size. The number of d.f. for the tests described here is always one less than the number of classes so that, for example, four classes have 3 d.f. In the present example (1), therefore, we have 1 d.f. and we use the top line in Table 4.3. We notice that our χ^2 value of 0.96 is less than 1.074 for P = 0.30 (or 30%), but more than 0.455 for P = 0.50 (or 50%). This tells us that if there were no real deviations from a 3:1 ratio, and we repeated the experiment a large number of times, deviations as great as, or greater than, the one we have here ($+/-6$) would occur in more than 30% of trials due to chance alone. In other words, there is a high probability that the deviation that we obtained in our experiment is due to chance, and we can confidently conclude that there is no significant departure from a 3:1 ratio. Our results agree with expectations.

The level of probability at which we decide to accept or reject our hypothesis of agreement with a given ratio is usually taken at 5% (P = 0.05). Any value of χ^2 smaller than 3.841, for one degree of freedom, is considered to be *non-significant* and the observed values are taken to agree with expectations.

Table 4.3 Distribution of χ^2

Degrees of freedom	Probability (P)								
	0.99	0.95	0.80	0.70	0.50	0.30	0.20	0.005	0.001
1	0.00016	0.004	0.064	0.148	0.455	1.074	1.6421	3.841	6.635
2	0.0201	0.103	0.446	0.713	1.386	2.408	3.219	5.991	9.210
3	0.115	0.352	1.005	1.424	2.366	3.665	4.642	7.815	11.341

Now consider the other set of data given in example (2) above, where we have red- and white-flowered classes of 165 and 35 respectively. The χ^2 value in this case works out at 6.0. For one degree of freedom, it falls between the probability levels of P = 0.01 (or 1%) and P = 0.05 (or 5%). Because P < 0.05, we conclude that the χ^2 value is *significant* at the 5% level. The probability that a deviation of this size is due to chance alone is very low – less than 1 in 20 – and we take this to mean that our hypothesis was incorrect, and that the data do *not* fit the expected 3:1 ratio.

As a third example, we will use Mendel's results for the testcross **RrYy** × **rryy** described on page 46. The hypothesis we wish to test is that the data agree with a $1:1:1:1$ ratio. Table 4.4 shows the observed results and the calculation of the χ^2 value.

Table 4.4

Phenotypic class	*O*	*E*	*d*	*d²/E*
'round, yellow'	55	52	+3	0.173
'round, green'	51	52	−1	0.019
'wrinkled, yellow'	49	52	−3	0.173
'wrinkled, green'	53	52	+1	0.019
	208	208	0	$\chi^2 = 0.384$

There are now four classes in our test and we use the row in Table 4.3 that corresponds to 3 d.f. Our χ^2 value falls between the probability levels of $P = 0.8$ (or 80%) and $P = 0.95$ (or 95%), and is therefore non-significant. There is a greater than 80% probability that the deviation from the expected values occurred by chance alone, and so we conclude that the data agree with our hypothesis.

The chi-squared test can be used to analyse any genetic ratios ($3:1$, $1:1$, $9:3:3:1$, $1:1:1:1$), including those that will appear in some of the forthcoming chapters (6, 7, 8 and 9). It is a valuable statistical device for assessing the reliability of data and for helping us to make decisions as to whether our data do, or do not, agree with a hypothesis that we have put forward. Where results do not conform to expectations it is then another matter to decide on the possible reasons for the lack of agreement.

Summary

◆ In Mendel's experiments, two or more pairs of contrasting characters were found to be inherited independently of one another. He concluded that this happened because each character was controlled by a different pair of cell elements, which we now call alleles of a gene.

◆ The two pairs of alleles segregate independently of one another so that the heterozygous F_1 dihybrid gives four kinds of gametes in equal frequencies.

◆ Random combinations of the gametes, from male and female F_1 individuals, gives a $9:3:3:1$ ratio of phenotypes in the F_2.

◆ The experiments on dihybrids gave rise to what is now known as Mendel's second law, or the law of independent segregation.

Further reading

1 H. Kalmus (1983) 'The scholastic origins of Mendel's concepts', *History of Science*, **21**, 61–83.
2 K. R. Lewis and B. John (1972) *The Matter of Mendelian Heredity*. Longman.
3 G. Mendel (1866) 'Versuche über Pflanzenhybriden', *Verhandlungen des Naturforschenden Vereins in Brüun*, **4**, 3–44.
4 G. Mendel (1965) *Experiments in Plant Hybridisation* (Mendel's original paper in English translation with Commentary and Assessment by R. A. Fisher; together with a reprint of W. Bateson's Biographical Notice of Mendel), edited by J. H. Bennett. Oliver and Boyd.
5 R. C. Olby (1966) *Origins of Mendelism*. Constable.
6 Vitezslav Orel (1996) *Gregor Mendel: The First Geneticist*. Oxford University Press.
7 R. E. Parker (1979) *Introductory Statistics for Biology*. Edward Arnold.
8 *MendelWeb* resource for teachers and students interested in classical genetics, introductory data analysis, and elementary plant science. Includes texts, illustrations, bibliographies and links to further resources. Mendel's original paper is given in German, with a revised version of the English translation: **http://www.netspace.org/MendelWeb**

Questions

1 State Mendel's first and second laws
  a as originally stated, and
  b using modern terminology.
  How would you explain these two laws? Do you think we are justified in calling them 'laws'?

2 A gene for coat colour in rabbits has four alleles. That for normal colour is **R** and is dominant to all the other alleles. The allele for 'albino' is **r**, and is recessive to all the others. The allele for 'Himalayan', r^h, is recessive to that for 'chinchilla', r^{ch}. What are the genotypes and phenotypes of offspring from a cross between homozygous 'chinchilla' males and heterozygous 'Himalayan' females?

3 In a particular variety of maize, the cobs have either red or purple grains. The colour is controlled by a single pair of alleles. The dominant allele **A** gives purple colour while the recessive allele **a** gives red colour.
  a State the genotype and phenotype of a heterozygous plant and a plant homozygous for allele **a**.
  b Grain colour is also affected by a second gene, which has alleles **E** and **e**. The gene is located on a separate chromosome. In the presence of allele **E,** colour develops normally – either red or purple depending on the genotype for the first gene. In plants homozygous recessive for this second gene (that is, **ee**) no colour develops regardless of the genotype for the first gene. Then the grains are not red or purple, but white. A plant of genotype **AAEE** is crossed with a plant of genotype **aaee**. What is the phenotype of these parent plants? What is the genotype and phenotype of offspring produced as a result of the cross (that is, the F_1)?
  c The F_1 from the cross are grown up and allowed to self-fertilise. Using a diagram, show the possible genotypes that will occur in the F_2 as a result of this cross. State the frequency of each genotype that would be expected among a large sample of F_2 progeny.
  d State the expected phenotypic ratio among the F_2.
  e Which genotypes would produce pure-breeding lines with white-grained cobs, if they were allowed to self-fertilise?

4 a Distinguish between the terms *gene* and *allele*.
  b In tomatoes, the gene for fruit colour has dominant allele **R** for red fruits, and recessive allele **r** for yellow fruits. Plant height is controlled by a gene with dominant allele **T** giving tall plants and recessive allele **t** for dwarf plants. One particular plant produces gametes of genotypes **RT, Rt, rT** and **rt** in roughly equal proportions. What is the genotype of this plant?
  c A second plant produces gametes of genotype **rt** and **rT** only. What is the genotype of this plant?
  d If the two plants were crossed what would be the expected genotypic and phenotypic ratios of the resulting offspring? Use a genetic diagram to explain your answer.

5 In guinea pigs, the gene for coat colour has allele **B** for black colour. Allele **B** is dominant over allele **b** for albino. A separate gene controls coat texture with the allele **R** for rough coat being dominant over the allele **r** for smooth coat. A heterozygous black, smooth-coated guinea pig is mated with an albino, heterozygous rough-coated guinea pig.

a What are the parents' genotypes? What are the genotypes of the gametes they produce?

b What are the genotypes of the offspring, and what phenotype will each genotype display?

c If the pair produced many offspring, or if many pairs of guinea pigs were mated with the same genotypes as those described in the cross above, what genotypic and phenotypic ratios would you expect to see among the young?

6 Pure breeding brown dogs have the genotype **aabb,** where the recessive allele **a** allows the development of coat pigment, and the recessive allele **b** is for an unlinked gene, and causes the pigment to be brown. The alternative allele for **a** is **A**, which prevents pigment production, resulting in white dogs. Allele **B** causes any pigment produced to be black. Pure-breeding brown dogs are crossed with dogs of genotype **AABB**.

a What is the phenotype of the dogs of genotype **AABB**?

b What is the genotype and phenotype of the puppies from the cross?

c The puppies are kept and eventually crossed with one another to produce F_2 offspring. Construct a Punnett square to show the possible genotypes of puppies in the F_2 generation.

d Given that the two genes are unlinked, what phenotypic ratio do you expect to find among the F_2?

e Of 200 F_2 puppies, 140 were white, 45 were black and 15 were brown. Use a chi-squared analysis to test whether the data fit the phenotypic ratio you have described. Show your working in tabular form.

f Dogs from the F_1 generation were backcrossed to the brown parental type. Show the genotypes that would be produced from this cross and state the phenotype for each of the genotypes.

7 A tall pea with red flowers was crossed to a tall pea with white flowers. The progeny had the following phenotypes: 91 tall red, 97 tall white, 29 dwarf red, 33 dwarf white. What were the genotypes of the parents?

8 The offspring of genetics breeding experiments may occur in certain well-defined ratios. Discuss the meaning of some of these ratios and their importance in determining the relative positions of genes.

9 Night blindness is a condition in which affected people have difficulty seeing in dim light. The allele for night blindness, **N**, is dominant to the allele for normal vision, **n**. (These alleles are *not* on the sex chromosomes.) The diagram shows part of a family tree showing the inheritance of night blindness.

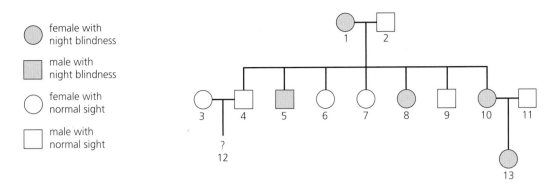

a Individual 12 is a boy. What is his phenotype?   [1]
b What is the genotype of individual 1? Explain the evidence for your answer.   [2]
c What is the probability that the next child born to individuals 10 and 11 will be a
   girl with night blindness? Show your working.   [2]

AEB A Biology Paper 1 June 1998 Q6

10 A series of breeding experiments was carried out with the fruit fly *Drosophila*.
   a A cross between pure-breeding males with red eyes and straight wings and females
      with purple eyes and curled wings produced offspring all with red eyes and straight
      wings. The same result was obtained when parents of the same pure-breeding
      phenotypes were used again but with the sexes reversed.
      (i) Suggest a reason for carrying out the cross with the sexes reversed.   [1]
      (ii) Using the letters **A** or **a** for eye colour and **B** or **b** for wing shape, state the
      genotypes of the parent flies in these crosses.   [2]
      (iii) State which of the phenotypes given above are controlled by dominant alleles.
      Give a reason for your choice.   [2]
   b Female offspring from the first experiment were then crossed with purple-eyed,
      curled-winged males. The progeny listed below were produced from this cross.

173 red eye, straight wing
165 red eye, curled wing
169 purple eye, straight wing
179 purple eye, curled wing

A null hypothesis was put forward that the genes for eye colour and wing shape are
not linked. A χ^2 test was carried out to test if this hypothesis was correct. Some steps
in the calculation of χ^2 are shown in the table below.

Phenotype	Observed number	Expected number	Difference $(O - E)$	$(O - E)^2/E$
red eye, straight wing	173	171.5	1.5	0.01
red eye, curled wing	165	171.5	−6.5	0.24
purple eye, straight wing	169	171.5	−2.5	
purple eye, curled wing	179	171.5	7.5	

(i) Copy and complete the table by calculating $(O - E)^2/E$ for the purple eye, straight wing flies and for the purple eye, curled wing flies. [2]

(ii) Using the formula below and the values from the table, calculate the value of χ^2. Show your working. [2]

$$\chi^2 = \Sigma \frac{(O - E)^2}{E}$$

(iii) State the number of degrees of freedom that should be used for reading the probability from the χ^2 table. [1]

London AS Biology and Human Biology AS Synoptic Paper Module Test B5/HB5

January 1998 Q2

Genes and chromosomes

Now that we have studied the way that the nucleus divides during cell division, and the way in which character differences are inherited in breeding experiments, we are in a position to see the parallel between these two aspects of the process of heredity. It is obvious to us now, as it was to Sutton in 1903, that if the genes are actually part of the chromosomes then the Mendelian principles of segregation and independent segregation can be readily explained by the way in which the chromosomes behave during the reduction division of meiosis. In this chapter, we discuss the relationship between genes and chromosomes and develop the idea of the **chromosome theory of heredity**. The theory will be summarised at the end of Chapter 6 after we have also dealt with the principle of 'linkage'.

Relationship between genes and chromosomes

The simplified diagram in Figure 5.1 shows the relationship between the genes and the chromosomes. In diploid organisms the chromosomes exist in homologous pairs. There are therefore two sets of chromosomes present in the nucleus of each cell, one set originating from the mother (maternal set) and one from the father (paternal set).

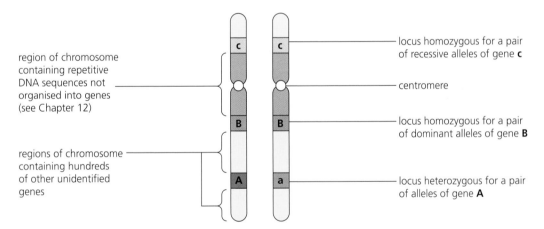

Figure 5.1 A pair of homologous chromosomes shown in an unreplicated form, as single chromatids, as they would be at the end of somatic cell division, or mitosis (see page 11). Three pairs of different genes are shown at different loci as well as some other features. A locus is the site in the chromosome where a gene is located: a particular gene is always found at the same locus. Homology means that a pair of chromosomes have alleles of the same genes, and other structural features, at corresponding places along their length

Each member of a homologous pair carries one of the alleles of each gene at corresponding positions called 'loci' (singular form is **locus**) along its length. In a pure-breeding homozygote the homologues carry identical dominant (**BB**), or recessive (**cc**), alleles at the loci concerned. In a heterozygote (**Aa**) they carry a dissimilar pair – each homologue carrying one of the two alleles.

Meiosis and segregation

Segregation of a pair of alleles (**Aa**) takes place at meiosis because the genes are part of the chromosomes and are separated from one another when the chromosomes move apart at the anaphase stages (see pages 18–19, 23). For a particular pair of alleles, the segregation may take place when the homologous chromosomes move to opposite poles at anaphase I, in which case it is referred to as 'first division segregation', or when chromatids are separated at anaphase II, when it is called 'second division segregation'. The difference between the two simply depends upon whether a cross-over takes place between the gene locus concerned and the centromere, as explained in Figure 5.2.

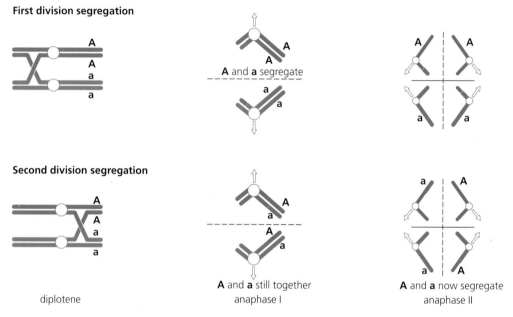

Figure 5.2 Chromosomal basis of segregation for a pair of alleles at a single locus. The diplotene bivalent is shown with a single chiasma. When there is *no* crossing over between the gene and the centromere the alleles segregate into different cells at AI (top). When crossing over does take place between the gene and the centromere segregation is delayed until the second division (bottom)

Normally, of course, we cannot tell at which stage of meiosis segregation takes place. We are unable to see the genes in the chromosomes and we have to deduce what is going on from our knowledge of meiosis and from the appearance of phenotypes in testcrosses and F_2 progeny. Ideally, we would like to look directly at the haploid cells, or gametes, coming from individual meioses – if we could find a suitable character. Ascomycetous fungi provide the right kind of material, and can be used to observe segregation in individual cells using spore colour mutants. The haploid products of each meiosis are the ascospores, which are held together in clusters within the ascus. These can be viewed with the aid of a microscope.

Meiosis and independent segregation

The independent segregation of two or more pairs of characters, according to Mendel's second law, takes place because the genes controlling the characters are located in different non-homologous chromosomes. The chromosomal basis of this second principle of genetics is simply the random way in which chromosomes attach to the spindle at the metaphase stages. An explanation of the mechanism is given for two pairs of alleles, **AaBb**, in the simplified diagram in Figure 5.3.

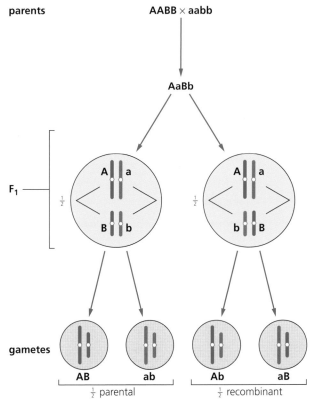

Figure 5.3 Simplified diagram explaining how the random orientation of bivalents on the spindle at MI of meiosis accounts for the independent segregation, and recombination, of pairs of genes on different chromosomes. Chromosomes are represented as single structures without showing their chromatids

The random orientation of bivalents with respect to one another leads to the production of four kinds of haploid gametes in equal frequencies. In dealing with independent segregation we do not need to concern ourselves with the consequences of crossing over within the individual bivalents. The effect of crossing over between the gene and the centromere is the same as described opposite for segregation, but the important factor here is the random orientation of the bivalents and it makes no difference to the overall outcome whether individual pairs of alleles segregate at AI or AII.

Independent segregation and recombination

The independent segregation of genes in different chromosomes leads to recombination. Of the four kinds of gametes produced, when two pairs of alleles are segregating, two of them (**AB**, **ab**) are parental and two (**Ab**, **aB**) are recombinants (Figure 5.3). This is known from the progeny phenotypes in the testcross, **AaBb** × **aabb**, and in the F$_2$ from the cross **AABB** × **aabb**.

When three pairs of genes are segregating independently, eight different combinations of the maternal and paternal chromosomes that carry them are found with equal frequency in the gametes. Two of the combinations are parental types and six are recombinant (Figure 5.4).

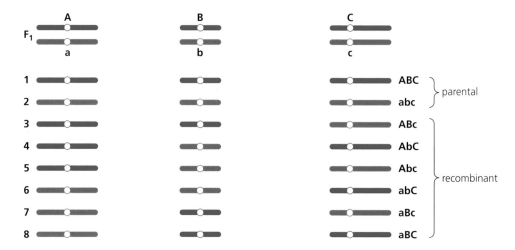

Figure 5.4

In general terms, and leaving aside any complications due to crossing over, the number of different combinations of maternal and paternal chromosomes that can occur in a sample of gametes by independent segregation is 2^n, where n = the number of chromosome pairs. For two chromosomes, shown in Figure 5.3, there are four possible kinds: $2^2 = 2 \times 2 = 4$. For three pairs, as in Figure 5.4, there are eight – that is, $2^3 = 2 \times 2 \times 2 = 8$ – and so on. Each additional pair of chromosomes *doubles* the number of combinations. The chromosome number of humans is $2n = 2x = 46$; that is, 23 pairs. Therefore, it is possible for an individual male to recombine his maternal and paternal chromosome sets, by independent segregation, to give $2^{23} = 8\,388\,608$ genetically unique sperms. Females have the same potential but they produce a relatively small number of egg cells during their lifetime.

We will discuss these matters of genetic variation further at the end of Chapter 6, and thereafter. For the time being, we will take note of the fact that the basic chromosome number of a species has some genetic significance. The larger the number of different chromosomes into which the genetic material is divided the greater are the possibilities for recombination by independent segregation.

Summary

◆ Genes are part of the chromosomes and the laws of heredity can be easily explained in terms of the way in which chromosomes behave when they undergo their reduction division at meiosis.

◆ Segregation is accounted for by the separation of homologous pairs at anaphase I, and by the movement apart of the daughter chromosomes at anaphase II when there has been a cross-over between the gene concerned and the centromere.

◆ Independent segregation is due to homologous pairs of chromosomes that orientate at random on the spindle at metaphase. Independent segregation gives rise to recombination and to genetic variation.

Further reading

The Natural History of Genes – Basic Genetics. Gives basic information on genes, chromosomes, DNA and inheritance:
http://glsc.genetics.utah.edu/basic/index/html

Questions

1 Explain the meaning of the term *homology*.

2 What is *independent segregation*?

3 Explain how first and second division segregations during meiosis lead to genetic variation.

Linkage and recombination

The characters that Mendel studied showed independent inheritance because they were determined by genes in different chromosomes. If all chromosomes were made up of only one gene, then the laws of segregation and independent segregation would provide us with an adequate description of heredity. But this is not the case. We now know that the chromosomes are actually made up of linear sequences of large numbers of genes, which are all *linked* together and which cannot behave independently of one another in their inheritance.

In this chapter, we will consider the inheritance of linked genes. We will also summarise the chromosome theory of heredity and the reasons why it is necessary to have meiosis at all.

Linkage

Linkage is the association of certain genes in their inheritance. The same term may also be used with reference to the characters determined by the linked genes. The concept of linkage can best be understood in the first place by describing an exceptional situation, and then proceeding to what is normal. In *Drosophila*, the males have an unusual meiosis. There is no crossing over between homologous pairs of chromosomes, and therefore no chiasmata are formed in the prophase stages. At metaphase I, the homologues simply lie side by side on the equator of the spindle and then separate from one another at anaphase I. Thereafter they behave normally. Because there is no exchange of chromatids between maternal and paternal partners in a bivalent, the genes in them show **complete linkage**.

Complete linkage

When pure-breeding male flies carrying the recessive mutations curled wings (**s**) and ebony body (**g**) are mated with normal females with straight wings (**S**) and grey body (**G**), the F$_1$ are all normal and heterozygous for both pairs of alleles. If the F$_1$ *males* (**SsGg**) are now testcrossed with double recessive females (**ssgg**), the four phenotypic classes that we might expect in the progeny, by independent segregation, are not present. Instead, there are only two phenotypic classes: those with the parental combinations of 'straight wings, grey body' and 'curled wings, ebony body'. The dominant and recessive alleles segregate from one another in the F$_1$ as completely linked pairs, and not independently. The reason is that the two genes are in the same chromosome (that is, the same pair of chromosomes) and not in two different ones as required for independent segregation. In the F$_1$, the two dominant alleles are together in the maternal chromosomes and the two recessive alleles are together in the paternal partner of the bivalent (Figure 6.1, overleaf).

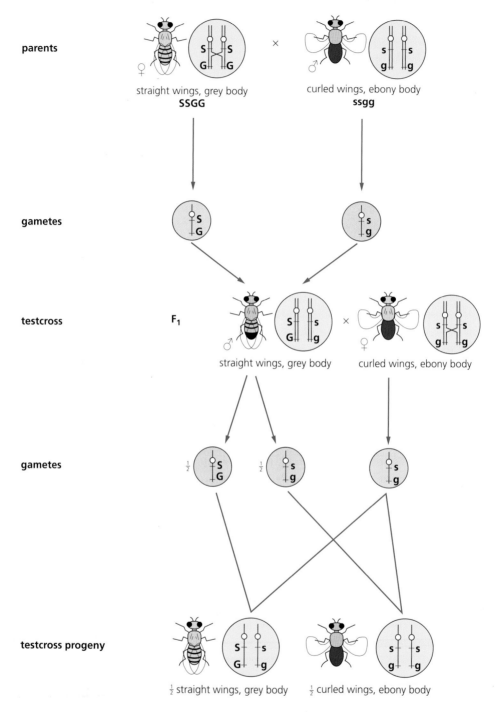

Figure 6.1 Complete linkage in the fruit fly, *Drosophila melanogaster*. In the F₁ hybrid males there is no crossing over and the paternal and maternal homologues segregate unchanged into the gametes, taking linked combinations of alleles with them. Consequently there are no recombinants and only two kinds of progeny are produced in equal numbers

In dealing with linkage, it is therefore helpful to change the notation slightly and to write the alleles one above the other, rather than side by side, and draw a line between them to represent the chromosomes (the alleles above the line are together on one chromosome, while those below the line are together on the other, Figure 6.2).

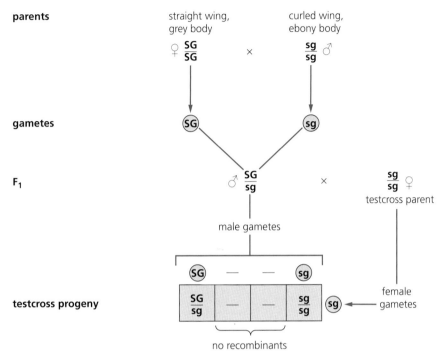

Figure 6.2

Since there is no crossing over in the male, and no exchange of chromatid segments between homologous partners, the genes are completely linked and simply follow the chromosomes in their inheritance (Figure 6.1). A bivalent without any crossing over behaves as in Figure 6.3.

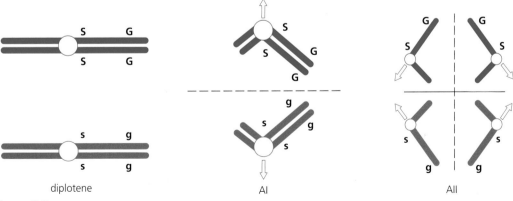

Figure 6.3

Partial linkage

If we consider the same cross again but this time using a female as the heterozygous F_1, and testcrossing her to a double recessive male, we get a different result (Figure 6.4).

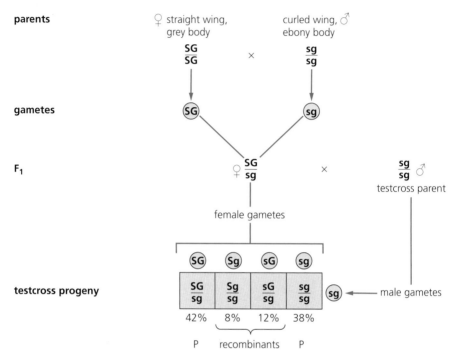

Figure 6.4

We now have four kinds of progeny resulting from the four possible combinations of alleles in the F_1 gametes. Two of them are 'parentals' (P), resembling the two parent flies going into the cross, and two are recombinants possessing characters of both parents. We notice that they are not in the equal frequencies expected by independent segregation, and neither are they completely linked, as they were when a male was used as the F_1 hybrid. This time we have **partial linkage** with the parental combinations making up the majority of the gametes produced by the F_1, and the minority being comprised of recombinants.

Crossing over

The reasons for this partial linkage have to do with the way in which the genes are arranged in the chromosomes and with the crossing over that takes place between them.

Crossing over is a process of exchange between homologous chromosomes, which gives rise to new combinations of characters. The explanation is given in the diagram in Figure 6.5. Crossing over gives rise to two kinds of recombinants in about equal numbers, because of the way in which two non-sister chromatids break and rejoin at corresponding sites within the bivalent during chiasma formation (Chapter 2).

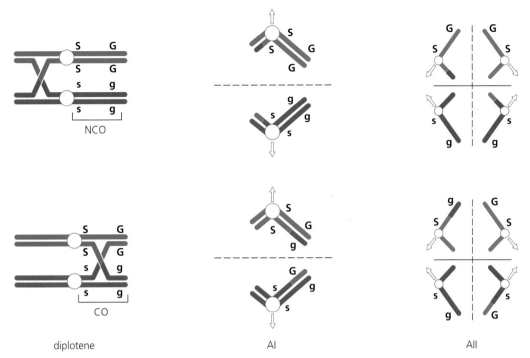

diplotene   AI   AII

Figure 6.5 Explanation of partial linkage in terms of chromosome behaviour at meiosis. Partially-linked genes are located in the same chromosome, and the parental combinations of alleles are transmitted together except in those cells where crossing over takes place between them. In a large sample of cells undergoing meiosis, crossing over will occur between the two genes in only some of the cells, depending on how closely linked they are. In cells in which crossing over does take place, only two of the four chromatids are involved. For these reasons recombinants in a testcross comprise less than 50% of the progeny. (NCO = no crossing over between the two genes; CO = crossing over)

When crossing over happens between the loci of genes we are studying, two of the four haploid products of meiosis are recombinant and two are parental. When crossing over doesn't happen between the gene loci, the products of meiosis are all parental. Since a cross-over can occur anywhere within a bivalent it will only take place between the two genes in some of the meiotic cells and not in others, and for this reason the recombinants usually make up much less than half of the testcross progeny. For any particular pair of genes, the proportion of recombinants will always be about the same when the testcross is repeated (around 20% for 'wing shape' and 'body colour' characters), because the genes occupy fixed positions in the chromosomes and there is a certain probability of a cross-over taking place between them. For a different pair of genes the proportion of recombinants will be different – as we will see later.

Crossing over and recombination

It is important to realise that crossing over is only one of the two ways in which recombination can occur. The other way is by the independent segregation of unlinked genes (Chapter 4).

Recombination and genetic variation

Significance of meiosis

In describing the principles of genetics – that is, segregation, independent segregation and linkage – we have covered the basic facts about inheritance and given the rules, or laws, by which the transmission of genetic information takes place. We have also seen how these principles can be explained and understood once we realise that the genes are parts of the chromosomes and are carried with them during their complex processes of pairing, recombination and separation at meiosis. But why is it all necessary? What is the significance of meiosis?

The answer is that during meiosis the genetic material in the maternal and paternal sets of chromosomes, which was originally brought together at fertilisation, is *recombined* and transmitted into the gametes in such a way that each sex cell produced has a unique set of genes (that is, combination of alleles). The progeny that are produced from crossing together two individuals of an outbreeding species therefore display genetic and phenotypic variability that results from the reshuffling of the genes in the sex cells of their parents. This reshuffling of genes is the function of meiosis and the advantage of sexual reproduction. The variation it generates is of fundamental importance to the long-term evolution of the species. Without genetic variability it would not be possible for natural selection to be effective and for a species to respond and to adapt, over generations, to changes arising in its environment (see Chapter 16).

The processes of recombination, as we have seen, are two-fold: (1) independent segregation of genes in different chromosomes; (2) crossing over between genes that are linked.

The F_1 from parents that differ in two genes produces four kinds of gametes. With independent segregation they will be in equal frequencies. With linkage they will not: there will be an excess of the two parental classes and a deficiency (< 50%) of the recombinant ones. If the genes are very closely linked, the recombinant gametes may be quite rare, but they will still be found if a large enough sample is taken.

In both cases, because four kinds of gametes are produced, the F_2 will comprise nine genotypes and four phenotypes (Figure 6.6).

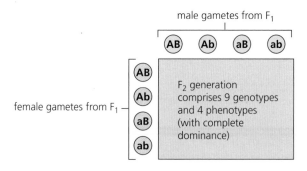

Figure 6.6 With independent segregation the phenotypes will be in the ratio 9 : 3 : 3 : 1. With linkage (that is, partial linkage) they will not be in any particular ratio, but they will all be present nonetheless

If the parents differed in three genes, the recombination in the F_1 would lead to an F_2 with 27 genotypes and eight phenotypes. Table 6.1 shows examples of what happens as the number of allelic pairs by which the parents differ is increased. The effectiveness of recombination in bringing about genetic variation is self-evident.

Table 6.1

No. of allelic pairs	F_2 genotypes	F_2 phenotypes*
2	9	4
3	27	8
4	81	16
10	59 049	1024
21	10 460 353 203	2 091 152
n	3^n	2^n

* where one allele is fully dominant over the other – this is not always the case, as explained in Chapter 8

In humans, it is estimated that there are of the order of 100 000 gene loci, distributed among the 23 chromosomes, and that on average about 7% of them are heterozygous. In other words an individual is heterozygous for some 6700 pairs of alleles and is therefore capable of producing 2^{6700} different kinds of germ cells. This is more than the number of atoms in the universe! Looked at in this way, it is not surprising that every single person in the world (except for monozygotic twins) is genetically unique.

Significance of diploidy

It follows from what has been said above that for meiosis to be effective in giving rise to genetic variation an organism must be sexually reproducing and have a diploid phase at some stage in its life cycle. Recombination can only occur when two sets of chromosomes that differ in the alleles of the genes they carry are brought together in the zygote during fertilisation. At meiosis, later on, the homologous sets will then pair and reshuffle their genes into the haploid spores or gametes that result from the reduction division process. Evolutionarily advanced species have, presumably, benefited from being diploid for the greater part of their life cycle and having a very short haploid phase. This results in their carrying many alleles hidden in the recessive form, which are of no obvious value or may even be harmful to the individual in the short term, but which provide a reservoir of genetic variation that is available for selective processes to act upon in the longer term. This aspect is discussed more fully in Chapters 15 and 16.

Significance of linkage

Genes in different chromosomes assort quite freely and segregate their alleles into the gametes in all possible random combinations. Linked genes assort much less freely. The linking together of certain combinations of genes into the same chromosome gives some control over recombination, and means that certain combinations of characters can be kept together.

Chromosome theory of heredity

The 'chromosome theory of heredity' states that the chromosomes are the carriers of the genes and represent the material basis of inheritance.

The idea was first put forward by Sutton in 1903 who saw that the separation of homologous chromosomes at anaphase could provide the material basis for the separation of character differences during gamete formation, as postulated by Mendel. Subsequently it was realised that chromosome behaviour during cell division could explain all the main aspects of 'nuclear inheritance' and the theory became firmly established. Now that we have dealt with the main aspects of inheritance, it is a useful stage at which to summarise the relationship between genes and chromosomes, on which the theory is founded.

1   Chromosomes contain the DNA, which is the genetic material. There is one double helix in each chromatid (Chapter 11). Duplication and separation of chromatids during the mitotic cycle accounts for the regular way in which identical sets of genes are distributed to the daughter cells in somatic cell division.
2   Segregation is explained by the separation of homologous pairs of chromosomes at anaphase of meiosis.
3   Independent segregation comes about due to the random orientation of bivalents on the spindle at metaphase.
4   Linked inheritance takes place for genes that are located in the same chromosome. Gene mapping has shown that there are as many groups of linked genes as there are numbers of chromosomes in the haploid complement of a species. *Drosophila* has four linkage groups for its four chromosomes. Maize has ten linkage groups corresponding to its basic number of $x = 10$.
5   Recombination of linked genes can be understood in terms of crossing over between chromatids of homologous partners.
6   Modified patterns of sex-linked inheritance are explicable in terms of the sex chromosome differences between males and females. In *Drosophila* and in mammals each gene in the differential segment of the X chromosome is represented by one allele in males and two in females. (This is explained fully in Chapter 7.)

Summary

◆ Genes that are located in the same chromosome show linked inheritance.

◆ In an F_1 hybrid the combinations of alleles present in the maternal and paternal chromosomes of a bivalent are transmitted together into the gametes, except in those cells in which they are recombined by crossing over.

◆ The incidence of crossing over between linked genes depends upon their placement in the chromosomes. Those that are far apart will have more crossing over between them than those that are close together because the occurrence of a cross-over in any particular site is a matter of chance.

◆ The correspondence between the number of linkage groups and the number of different chromosomes, for a particular species, provides additional evidence for the chromosome theory of heredity.

◆ Crossing over between linked genes is one of the ways in which recombination takes place at meiosis. The other way is independent segregation.

◆ The variation that comes about by recombining heterozygous combinations of alleles is the main function of meiosis and of the sexual system of reproduction.

Further reading

Virtual FlyLab. Teaches the principles of genetic inheritance using virtual experiments: **http://biologylab.awlonline.com/**

Questions

1 Explain what is meant by the terms *gene linkage* and *recombination*.

2 Explain why no two sperms produced by a man are of the same genotype.

3 Outline the main events of meiosis. Explain how meiosis generates variation.

4 Explain the term *cross-over frequency*. The cross-over frequencies between four genes were determined as follows: 20% between **A** and **B**, 5% between **A** and **C**, 5% between **B** and **D** and 30% between **C** and **D**. What is the probable sequence of these genes on the chromosome? What would you expect the cross-over frequency between **A** and **D** to be? Discuss the significance of crossing over.

5 Bateson and Punnett studied the genetics of sweet pea plants. They crossed plants with red flowers and round pollen grains with plants with purple flowers and long pollen grains. All the F_1 plants had purple flowers and long pollen grains. The F_2 results are shown in the table below.

a Copy and complete the table, showing the expected proportions and numbers of plants if the two genes are unlinked.

Flower colour	Pollen grain shape	Number of plants in F_2	Expected proportion	Expected number
purple	long	4831		
purple	round	390		
red	long	393		
red	round	1338		

b How would you explain the observed ratio of approximately $11:1:1:3$?

c By what process have the new combinations of characters, such as red flowers and long pollen grains, come about?

6 In the fruit fly *Drosophila* the genes for 'normal antennae' and 'grey body' are linked and are dominant to 'twisted antennae' and 'black body'. When a normal fly was crossed with one carrying recessive alleles the offspring were all of the normal type. One of the offspring was crossed to a fly homozygous for the recessive alleles. The following offspring were obtained.

Antennae	Body colour	Number of flies
normal	grey	90
normal	black	9
twisted	grey	11
twisted	black	90

Explain this result and make an annotated diagram of the bivalent that produced the recombinant classes. State the recombination frequency for the two genes.

7 Give an illustrated account of meiosis and genetic variation.

7　Sex-linked inheritance

Sexual differentiation of a species into separate male and female forms is usually genetically determined. Geneticists soon realised that the inheritance of sex could be interpreted in Mendelian terms, since a mating between male and female always resulted in two sexes among the offspring in approximately equal numbers – as would be achieved in a monohybrid testcross. In some species sexuality is simply associated with a single gene difference; for example, in the mosquito the male is **Mm** and the female **mm**. More commonly, however, an examination of males and females reveals a difference in their chromosome complements, and sex determination is associated with the inheritance of particular chromosomes.

The term **sex chromosomes** is used to describe the chromosomes in the complement that carry the sex-determining genes. All other chromosomes in the set that are not involved in sex determination are known as **autosomes**.

As the sex chromosomes are usually different in structure and genetic organisation in the two sexes, there are modified patterns of inheritance of the genes that are located in them. Indeed, the association of modified patterns of inheritance with a particular pair of chromosomes within the complement provided some of the earliest and most convincing evidence for the chromosome theory of heredity.

In this chapter, we consider some of the well-known sex chromosome systems in animals, and the modified forms of Mendelian inheritance that are linked to them. We also discuss mechanisms of sex determination.

Sex chromosomes

The XO system

Early cytologists studying chromosomes in insects belonging to the order Orthoptera (grasshoppers and locusts) found that one member of the chromosome complement, which was paired in the female, had only a single representative in the male. This became known as the X chromosome. In the grasshopper *Chorthippus parallelus*, for example, the male has 17 chromosomes (including one X) and the female 18 (including two Xs). During meiosis in the female the two Xs pair together and behave in the same way as any other bivalent. In the male, the single X has no pairing partner: it moves undivided to one pole of the cell at anaphase I and then divides normally at AII. Half of the sperms therefore contain a single X and the other half do not (O).

Since the male produces two types of gametes with respect to the sex chromosomes, it is referred to as the **heterogametic sex** (XO). The female (XX) produces only one kind of gamete (containing an X) and is therefore called the **homogametic sex**.

The transmission of the sex chromosomes follows a simple pattern of inheritance, and the sex ratio of $\frac{1}{2}$ male : $\frac{1}{2}$ female is maintained in the progeny of crosses (Figure 7.1, overleaf).

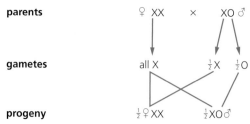

Figure 7.1 Transmission of sex chromosomes in the grasshopper *Chorthippus parallelus*

The XX/XY system

In many species of flies, including *Drosophila*, and in many mammals, including humans, there are two kinds of sex chromosomes. They are designated X and Y. Their morphology and relative sizes vary according to the species. The X may be the largest or, less commonly, the smallest member of the complement. It is usually larger than the Y, as is the case in humans (Figure 7.2), but is occasionally smaller, as in *Drosophila* (Figure 7.3).

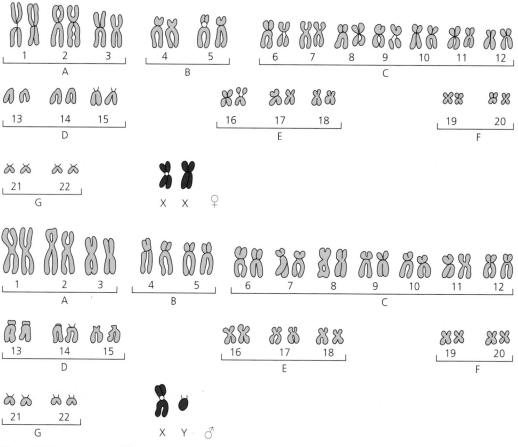

Figure 7.2 Karyotypes of humans, showing the difference in chromosome complements between sexes. The X and Y are the sex chromosomes and pairs 1–22 are the autosomes

In flies and in mammals the female is the homogametic sex (XX), producing eggs carrying a single X, and the male is heterogametic (XY). At meiosis in the male, the X and Y form a bivalent in prophase I and then segregate to opposite poles of the cell at anaphase I. At anaphase II, the chromatids separate so that half of the sperms contain a single X and the other half a Y. The sex of the offspring is therefore determined by the male parent, as illustrated in Figure 7.3 for *Drosophila*. In a large sample of offspring half will be male and half female. For an individual, the probability of being male or female is $\frac{1}{2}$ in both cases.

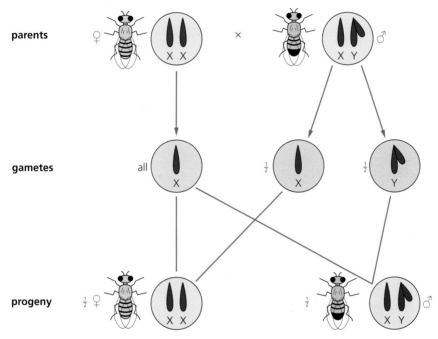

Figure 7.3

In butterflies, moths and birds, the same sex chromosome system operates but the roles are reversed: the homogametic sex is the male and the heterogametic sex is the female.

Many different sex-determining systems operate among higher organisms. An interesting example is that found in the honeybee. In this species, sex is determined by the ploidy level (Chapter 14), which is controlled by the female queen bee. The queen is diploid with $2n = 2x = 32$. By controlling the sphincter of her sperm receptacle she can lay either unfertilised eggs with the haploid (reduced) number of 16 chromosomes, or fertilised eggs with 32 chromosomes. The fertilised diploid eggs develop into females, and the way in which each is fed determines whether it will develop into a worker or a queen. Unfertilised haploid eggs become males (that is, drones). Since the male is haploid, with only one set of chromosomes, sperm production depends on a modified meiosis that is essentially a mitosis, thus avoiding further reduction in chromosome number.

Sex-linked genes

Sex chromosomes also carry genes that are not directly involved with the determination of sex. The characters that these genes control are associated in their inheritance with the character of sex. For this reason genes that are located in the sex chromosomes, and which are linked to the sex-determining genes, are called **sex-linked genes**.

The inheritance of sex-linked genes is different from that for autosomal genes, because in the heterogametic sex the sex chromosomes are not completely homologous with one another. In the XX/XY system, with which we are mainly concerned, there are certain genes in the differential part of the X chromosome that have no allelic partners in the Y, and vice versa. A general scheme for the structure and genetic organisation of the XY sex chromosome pair is shown in Figure 7.4.

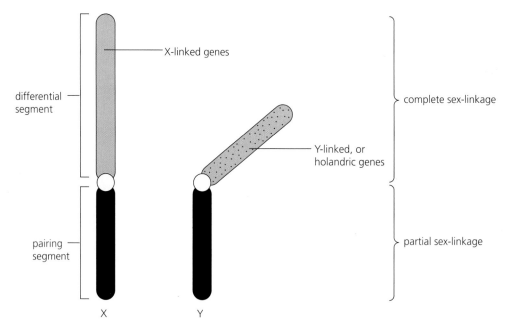

Figure 7.4

Genes in the homologous pairing segment, in which there may be crossing over between the X and Y, show *partial* sex-linkage. They have normal diploid inheritance like those in the autosomes. Their presence in the sex chromosomes is detected because they show linkage to the genes in the differential segment, which are the ones with the distinctive patterns of sex-linked inheritance. Genes in the differential segment have no crossing over between them and are said to show *complete* linkage. As it happens, the Y chromosome in most species is largely devoid of genes, as in *Drosophila* and humans, and the pairing segment is often quite small. In talking about sex-linked inheritance, therefore, we are mainly concerned with the genes in the differential segment of the X chromosome.

Inheritance of sex-linked genes

What is the pattern of inheritance of sex-linked (that is, X-linked) genes, and how do we detect such genes?

Eye colour in *Drosophila*

The first evidence of sex-linkage came in 1910 from an experiment by Morgan and his co-workers using *Drosophila*. Among his cultures of normal red-eyed flies Morgan found a white-eyed male. He isolated this male and crossed it with its red-eyed sisters. He found that all the F_1 had red eyes, showing that 'white eye' was recessive. When the F_1 were mated among themselves he found red-eyed and white-eyed flies in a 3:1 ratio in the F_2, but all the white-eyed flies were *male*.

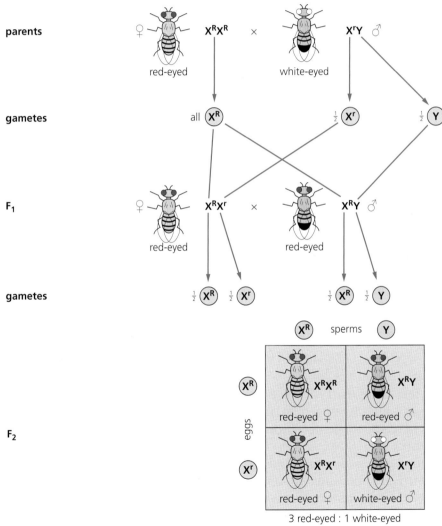

Figure 7.5

Morgan was puzzled by the absence of white-eyed females in the F_2. He found that if he took the red-eyed F_1 females and backcrossed them with the white-eyed male parent, the testcross progeny comprised white-eyed and red-eyed males and females in a $1:1$ ratio. This showed that it *was possible* for a female to have white eyes and also that the female F_1 progeny *were* heterozygous in genotype. How then can we explain the absence of white-eyed females in the original F_2? The explanation arises because the gene concerned is located in the X chromosome; that is, it is X-linked. We can represent the cross using the notation X^R and X^r to indicate the location of the gene on the X, and Y to indicate the Y chromosome (Figure 7.5).

Once the gene is located on the X, we can see how the results are obtained. The female in *Drosophila* is the homogametic sex, and so for the gene concerned she can be:

- $X^R X^R$ homozygous red-eyed
- $X^R X^r$ heterozygous red-eyed
- $X^r X^r$ homozygous white-eyed.

The male, however, only carries one X chromosome and has no corresponding genes on the Y. As a consequence, whatever is carried on the X chromosome is obvious in the phenotype of the male, because it is present singly, and there is no other member of the chromosome complement whose genotype can mask it. Genes present only once in the genotype, and not in the form of pairs, are said to be **hemizygous**. The male can therefore be:

- $X^R Y$ hemizygous red-eyed
- $X^r Y$ hemizygous white-eyed.

The female parent was homozygous red-eyed and produced only one kind of gamete, X^R. The male was hemizygous white-eyed and produced two kinds of gamete, X^r and Y. In the F_1, both male and female are red-eyed but the males are hemizygous ($X^R Y$) and the females heterozygous ($X^R X^r$). In the F_2, half of the male progeny received X^R from the female (and Y from the male) and half received X^r. Because they are hemizygous, the genotype is apparent in the phenotype and half the males are red-eyed and half white-eyed. The female F_2 progeny receive X^R from the male and either X^R or X^r from the female. They are therefore all red-eyed.

In this experiment sex-linkage was indicated because of the difference between males and females in the F_2. Females were all red-eyed, while half the males were red-eyed and half were white-eyed. In all other monohybrid crosses we have examined so far, which were not concerned with sex-linked genes, the pattern of inheritance was independent of sex.

The second indication of sex-linkage comes when reciprocal crosses are made. When the female is now used as the white-eyed parent, and the male as the red-eyed parent, there is a different outcome to that shown in Figure 7.5. Both red-eyed and white-eyed flies are found in the F_1 and there is now a $1:1$ ratio of red:white in the F_2 (Figure 7.6).

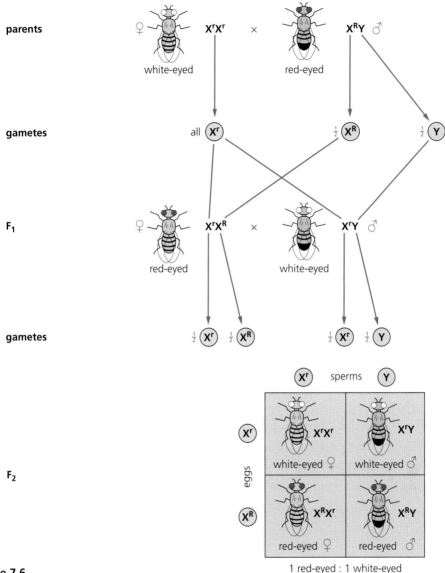

Figure 7.6

We notice that in this second cross the females in the F_1 resemble the eye-colour character of their fathers, and the males resemble that of their mothers. This pattern is referred to as **criss-cross inheritance** and is another indication that a character is controlled by a sex-linked gene.

Sex-linkage in humans

A number of human characters appear much more frequently in males than they do in females, and show a pattern of inheritance that is consistent with sex-linkage. The best known ones are haemophilia and colour blindness, described overleaf.

Haemophilia

Haemophilia is a disease of the blood. The most common form is caused by a single recessive gene. Its main feature is reduced ability of the blood to clot so that haemophiliacs are susceptible to protracted bleeding, even from trivial wounds. Haemophilia has been carefully pedigreed in the royal family of Queen Victoria. The lineage is shown in Figure 7.7.

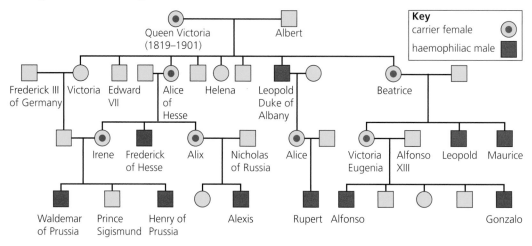

Figure 7.7 Pedigree of haemophilia in the royal families of Europe (excluding many individuals from later generations)

The fact that only male members of the family showed the disease is indicative that this is a sex-linked character. In humans, we recall, the female is the homogametic sex and so can be:

- $X^H X^H$ normal, homozygous
- $X^h X^h$ haemophiliac, homozygous
- $X^H X^h$ normal, heterozygous 'carrier'

The male, on the other hand, is heterogametic and does not carry the haemophilia locus on the Y chromosome. Males can therefore be:

- $X^H Y$ normal, hemizygous
- $X^h Y$ haemophiliac, hemizygous

Since one of Queen Victoria's sons was a haemophiliac and Albert of Saxe-Coburg-Gotha (her husband) was normal, it is clear that Queen Victoria must have been a carrier. We can represent the marriage as shown in Figure 7.8.

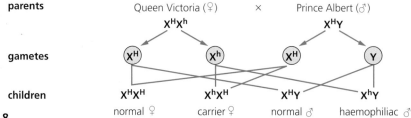

Figure 7.8

Queen Victoria produces two types of gametes, half of which carry **H** on the X chromosome and half carry **h**. On average, therefore, we would expect that half her sons would receive X^H (and be normal X^HY), and half would receive X^h (and be haemophiliac X^hY). Her daughters receive an X with the normal **H** allele from Prince Albert, and either X^H or X^h from Queen Victoria. They will therefore be either carriers, or homozygous normal. Some of her daughters were clearly carriers, since haemophiliac males appear among their children, and in total three generations were affected by the disease. Female haemophiliacs are only rarely found because they depend on sexual partnerships between haemophiliac males and carrier (or haemophiliac) females, which do not occur very often.

Colour blindness

In humans, one form of colour blindness results from a recessive gene carried on the X chromosome. Now consider the following questions.

A colour blind boy had parents with normal colour vision.
(1) What was the genotype of the mother? What was the genotype of the father?

Since both parents are normal, the mother must have been a carrier. Using **C** for the normal allele of the gene and **c** for the allele causing colour blindness, we can represent the mother as X^CX^c. The father was also normal and must have been X^CY.

(2) If the boy's brother, who has normal colour vision, marries a girl who is heterozygous for colour blindness, what would be the possible phenotypes and genotypes of their children?

The boy's brother is normal and so must be X^CY, like his father. The phenotypes and genotypes of their children would be as in Figure 7.9.

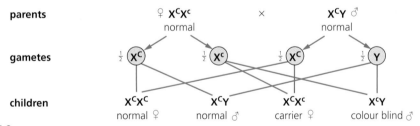

Figure 7.9

Duchenne Muscular Dystrophy

One of the most serious X-linked recessive genetic disorders is Duchenne Muscular Dystrophy (DMD). It occurs at a frequency of about 1 in 3500 male births due to spontaneous mutation (Chapter 13). Although it is on the X chromosome, the condition rarely occurs in females as the males do not live long enough to pass the allele on to daughters, so there are no female carriers. Babies born with DMD appear to be fine until they reach about two to four years old. Their skeletal muscles do not then develop properly and are weak. By the age of 10–12 children with DMD are often wheelchair-bound, and they usually die sometime between their mid-teens and early adulthood, often of respiratory failure. At the moment there is no cure for the disease, and little treatment for alleviating the condition of sufferers.

X-linked mental disorders

Some degree of mental impairment affects perhaps 2–3% of the population, the degree varying from very mild to very severe. Although mental retardation may be associated with Down's syndrome or metabolic disorders, such as phenylketonuria, the cause is often unknown. More males are affected than females, and several genes associated with mental retardation are on the X chromosome. The most common of these is 'fragile-X syndrome', or 'Martin–Bell syndrome' after the man who first described it in 1943. It is thought that fragile-X syndrome is second only to Down's syndrome as a genetic cause of mental handicap. The degree of mental retardation associated with the syndrome can be from mild to severe. The symptoms vary between individuals, but can include speech and learning disabilities, hyperactivity, autism, heart ventricular disease and facial dysmorphism. The syndrome is caused by a dominant mutation, and occurs in about 1 in 3000 children.

Other X-linked genes

Other X-linked genes in humans include 'Christmas disease', which is named after the patient in whom it was first described. This is another type of haemophilia caused by reduction in the amount of plasma thromboplastin component (PTC), and is called haemophilia B to distinguish it from haemophilia A (the type in Queen Victoria's family), which results from reduction in the amount of anti-haemophiliac factor (AHF).

Y-linked genes in humans

There are very few known examples of Y-borne genes in humans. This is surprising since it should, in theory, be easy to detect the effect of Y-linked genes in men, when there are no homologous genes (alleles) on the X chromosome to mask their effects. Some pedigrees suggesting Y-linkage have been reported, including one for a defect described as 'porcupine skin' and one for webbed toes, but fuller examination raised scepticism as to their validity.

A Y-linked gene can only ever be found in males and a man will pass his Y-linked genes to all his sons. A possible example is that of the character 'hairy ears', which has been traced over seven generations in one particular pedigree and was only shown by the male members. Also a gene influencing the development of male sexual characteristics has recently been shown to be on the Y chromosome. The gene, called SRY (for sex-determining region Y), codes for a testis-determining factor that initiates testis formation in embryos. An interesting aspect of this gene is that when the small piece of the Y chromosome that carries it is translocated into an X chromosome it can make what should be an XX female into a male. Although new molecular genetics techniques will allow the identification of the genes on the Y chromosome it is likely that there will be fewer genes on this chromosome than on any other of the human chromosomes.

Summary

◆ Sexual differentiation into male and female forms is mainly determined by particular chromosomes within the complement, called sex chromosomes.

◆ These differ in their form, and gene content, and in the ways in which they bring about the determination of sex.

◆ Genes located in the sex chromosomes show modified patterns of inheritance known as sex-linkage, because the characters they determine are associated in their inheritance with sex.

Questions

1 a Explain what is meant by the term *sex-linkage*. Use examples to illustrate your answer.

  b Why can a man not pass on an X-linked gene to his sons?

2 Describe the different mechanisms by which sex may be determined in animals.

3 Explain what is meant by the terms *complete linkage*, *partial linkage* and *sex-linkage*. How does linkage cause variations in Mendelian inheritance?

4 The gene controlling barring on the plumage of chickens is sex-linked and dominant. It can be used to sex newly hatched chicks in crosses between barred hens and non-barred cockerels. What is the phenotype of female chicks? Construct a diagram to show how this cross works.

5 The ability to taste phenyl thiourea is determined by a dominant allele of an autosomal gene. One form of red–green colour blindness is determined by a recessive allele of a sex-linked gene. A husband and wife are both 'tasters' with normal vision.

  a What is the expected proportion of children with the same phenotype as the parents?

  b The couple have a 'non-taster' child who is colour blind. What sex is the child?

  c What is the expected proportion of children with this phenotype?

6 In *Drosophila*, 'white eye' (**r**) is recessive to the normal 'red eye' (**R**). The gene is sex-linked. What genotypes and phenotypes would you expect among the offspring of a cross between red-eyed males and white-eyed females? If the F_1 were allowed to interbreed what genotypes and phenotypes would you expect among the F_2? If virgin F_1 females had been isolated and then crossed with a red-eyed male, what genotypes and phenotypes would you expect to find among their progeny?

7 In mice, the gene for albinism is recessive and autosomal, that for coat dappling is dominant and sex-linked. (A coat can only be dappled if it is not also albino.) A dappled male mouse is mated with a plain-coated non-albino female. Some of the offspring are albino.

  a Is it possible to distinguish offspring sexes by the phenotype of the coat? Why?

  b What proportion of the offspring are expected to be albino? Illustrate the cross using a Punnett square.

8 In cats, one of the genes for coat colour is present only on the X chromosome. This gene has two alleles. The allele for ginger fur, X^B, is dominant to that for black fur, X^b.

  a All the cells in the body of a female mammal carry two X chromosomes. During an early stage of development one of these becomes inactive and is not expressed. Therefore, female mammals have patches of cells with one X chromosome expressed and patches of cells with the other X chromosome expressed. Tortoiseshell cats have coats with patches of ginger and patches of black fur.

   (i) What is the genotype of a tortoiseshell cat? [1]

   (ii) Explain why there are no male tortoiseshell cats. [1]

b A cat breeder who wishes to produce tortoiseshell cats crossed a black female cat with a ginger male. Copy and complete the genetic diagram and predict the percentage of tortoiseshell kittens expected from this cross.

parental phenotypes	black female	×	ginger male

parental genotypes   _____   _____

gamete genotypes   _____   _____

offspring genotypes   _____

percentage of tortoiseshell kittens   _____

AQA (AEB) AS/A Biology Module Paper 2 June 1999 Q3

9 In the fruit fly, *Drosophila melanogaster*, the two loci controlling red or white eye colour and grey or yellow body colour are on the X chromosome. The allele, **R**, for red eye colour is dominant to that for white, **r**, and the allele, **G**, for grey body colour is dominant to that for yellow, **g**. In *Drosophila*, male flies have the genotype XY.

   Homozygous white eye, yellow body female flies were cross with red eye, grey body males, to give an F_1 generation.

a Draw a genetic diagram to show the genotypes of the parents and the genotypes and phenotypes of the F_1 generation of this cross. [4]

The F_1 flies were interbred to give an F_2 generation, and the phenotypes of the F_2 flies examined. The numbers of male flies and their phenotypes are shown in the table.

Phenotype		Number of F_2 flies	
Eye colour	**Body colour**	**Male**	**Female**
white	yellow	237	
red	grey	256	
white	grey	3	
red	yellow	4	
		Total 500	

b (i) Copy and complete the table to show the probable number of female flies showing each phenotype. [2]

(ii) Explain the results obtained in the F_2 generation. [4]

UCLES A Modular Sciences: Applications of Genetics March 1998 Section A Q2

10 Tongue-rolling and red–green colour blindness are two genetically controlled conditions which occur in humans. Tongue-rolling is controlled by the dominant allele, **T**, while non-rolling is controlled by the recessive allele, **t**. Red–green colour blindness is controlled by a sex-linked gene on the X chromosome. Normal colour vision is controlled by the dominant allele, **B**, while red–green colour blindness is controlled by the recessive allele, **b**.

a Copy and complete the genetic diagram to show the possible genotypes and phenotypes which could be produced from the following parents. [4]

parental phenotypes	female colour blind and heterozygous for tongue-rolling	×	male normal vision and non-roller
parental genotypes	_____		_____
genotypes of gametes	_____		_____
genotypes of children	_____		
sex and phenotypes of children	_____		

b Explain why a higher percentage of males than females in a population is red–green colour blind. [1]

c Sex-linked genes on the Y chromosome have been found in humans and other animal species. Suggest and explain *one* piece of evidence which would support the presence of such a gene. [1]

AQA (NEAB) A Biology Paper 1 June 1999 Section B Q16

Modified genetic ratios

Crosses between pure-breeding individuals that differ in one or two pairs of genes do not always give the classical 3:1 and 9:3:3:1 Mendelian ratios in the F_2. This may be because the genes are linked (Chapter 6), or sex-linked (Chapter 7). Genetic ratios depend, however, on the way in which genes are *expressed* in the phenotype, as well as on their mode of *transmission*. So far we have dealt with simple examples of gene action. For a single character, determined by one gene, one of the alleles has always shown complete dominance over its partner in the heterozygote. Where two independently segregating genes (**A** and **B**) are involved, they act separately from one another and control two different characters. Gene action is not always this straightforward.

Alleles of a single gene may interact together and give rise to F_1 phenotypes that are dissimilar to both of the two parents, and to F_2 progeny that have a modified ratio of phenotypes. Likewise two independently-inherited genes may both affect the same single character and this again will lead to modifications of the familiar 9:3:3:1 ratio in the F_2.

Interaction between alleles

Incomplete dominance

In cases of **complete** (or simple) **dominance** the heterozygote (**Aa**) has the same phenotype as the dominant homozygote (**AA**). This is the relationship between a pair of alleles that we have been concerned with in the previous chapters – because it provides the simplest situation for study. With **incomplete dominance** (semi-dominance, partial dominance) the heterozygote exhibits a phenotype that is *intermediate* between the two homozygous forms. An example is found in the four-o'clock plant, *Mirabilis jalapa*. When pure-breeding red-flowered four-o'clocks are crossed to pure-breeding white-flowered ones, all the F_1 have pink flowers. Selfing the F_1 gives a ratio of 1 red:2 pink:1 white among the F_2. The cross may be represented as shown in Figure 8.1.

The genotypic ratio is exactly what we would expect for a monohybrid cross, but the phenotypic ratio is modified to 1:2:1, and is the same as the genotypic ratio, because incomplete dominance results in the heterozygotes being intermediate between the two homozygous forms.

Another example of incomplete dominance occurs in domestic fowl. In pure-breeding Andalusians, black feather colour is due to the formation of the pigment melanin. A pure white strain was found that was lacking in melanin. The F_1 between the two strains has an intermediate colour called 'slate blue', which is due to partial development of melanin. Crossing F_1 birds gives a 1:2:1 ratio of black:blue:white in the F_2.

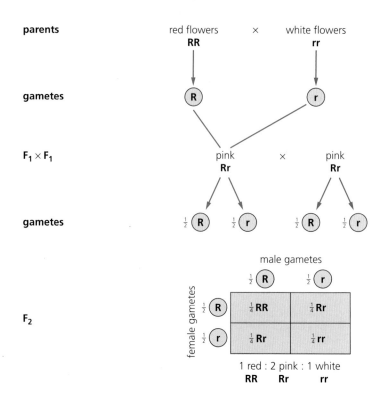

Figure 8.1

It was probably these kinds of observations that led to the earlier idea of 'blending inheritance', which we now know to be incorrect (Chapter 4). The point is that although the F_1 are intermediate in character, the parental types segregate out again in the F_2. This shows that the alleles themselves have not 'blended' together in the F_1, but have remained separate from one another as discrete units of inheritance. It is their *interaction*, at the level of gene action and expression, that gives rise to the intermediate phenotype.

Co-dominance

In **co-dominance** both alleles are expressed in the phenotype and the heterozygote has the characters of both parents.

The MN blood group system in humans provides us with an excellent example. Antigenic proteins on the surface of the red blood cells give individuals their blood group phenotype. There are three blood groups in this system – M, N and MN – determined by two co-dominant alleles of a single gene. People who are homozygous **MM** have blood group M and produce only M antigens on their red cells; **NN** homozygotes are group N, with N antigens; and heterozygotes, **MN**, are group MN and produce both M and N antigens. In a cross **MM × NN,** the F_1 (**MN**) have the characters of both parents; that is, they produce both kinds of antigens. In the F_2, there is a modified $1:2:1$ phenotype ratio of 1 M:2 MN:1 N due to co-dominance. The blood group genetics of humans is discussed in more detail in Chapter 9.

Lethal alleles

Lethal alleles are not actually a form of allele interaction, but since they also lead to modified F$_2$ ratios they are dealt with in this section.

In 1905 the French geneticist Lucien Cuénot studied a strain of mice with yellow fur instead of the normal grey (agouti) fur. On crossing 'yellow' with pure-breeding 'grey' he obtained a 1:1 ratio of 'yellow':'grey' among the progeny, indicating that yellow mice were heterozygous (**Yy**) and that 'yellow' was dominant. When two 'yellows' were mated together the progeny were in the ratio of 2 'yellow':1 'grey', which was a puzzling deviation from the expected 3:1 ratio. The clue to this deviation emerged when it was realised that pure-breeding homozygous 'yellows' (**YY**) were never found, and that all yellow mice were heterozygotes. Cuénot concluded that the gene for 'yellow' must be lethal in the homozygous form. This was confirmed by finding aborted fetuses in **Yy** females. The modified 2:1 ratio is therefore explained (Figure 8.2).

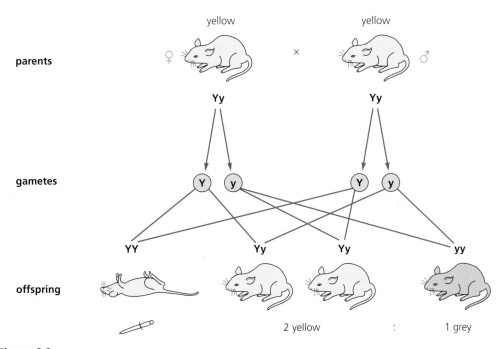

Figure 8.2

An interesting aspect of gene action in this example is the way in which allele **Y** is dominant in its effect on coat colour, because the heterozygotes (**Yy**) are yellow, and recessive in terms of its effect on lethality – since it only results in death when present in the homozygous form. We therefore say that **Y** is a dominant visible and a recessive lethal gene.

Common forms of lethal genes in plants are those that cause a lack of chlorophyll. These are usually recessive mutations and death occurs in the seedling stage because homozygous recessives are unable to engage in photosynthesis.

Interaction between genes

Epistasis

In **epistasis** (which means 'standing upon'), one gene *hides* the expression of another one. Epistasis is involved in many kinds of gene interactions.

Coat colour in mice, and in domestic fowl

A well-known example of epistatic gene action is that of coat colour in mice. Wild mice have agouti fur. The individual hairs are black with a yellow band, and this combination of colours produces the typical mouse 'agouti' phenotype. 'Non-agouti' mice have black coats, composed of uniformly black hairs. The difference is due to a single gene with two alleles, **A** and **a**. 'Agouti' (**AA, Aa**) is dominant over 'black' (**aa**). A second independently-inherited gene is required for the synthesis of the hair pigment. When at least one dominant allele of this gene is present (**CC, Cc**) the 'agouti' and 'black' phenotypes can be expressed. When this second gene is in the homozygous recessive form (**cc**) no pigment is formed and the mice are albino.

The gene for pigment development is therefore epistatic, when homozygous recessive, over both 'agouti' (**AA, Aa**) and 'black' (**aa**).

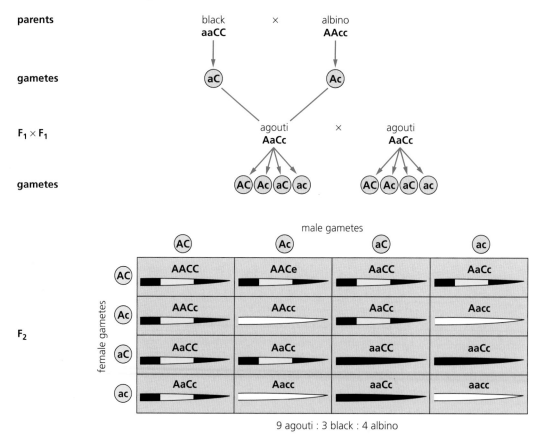

9 agouti : 3 black : 4 albino

Figure 8.3

In a cross of pure-breeding black (**aaCC**) × pure-breeding albino (**AAcc**) mice, the F_1 are all agouti. Interbreeding the F_1 produces a modified F_2 ratio of 9 agouti : 3 black : 4 albino. The $\frac{9}{16}$ that are agouti have at least one dominant allele of both genes. The $\frac{3}{16}$ that are black are homozygous recessive for **a** (**aa**) and have at least one dominant allele of gene **C**. The $\frac{4}{16}$ that are albino are the result of the epistatic action of the recessive homozygote, **cc** (Figure 8.3).

Epistasis can also be seen in the inheritance of plumage colour in domestic fowl. Pure-breeding 'white leghorns' are homozygous dominant for a gene that causes inhibition of colour in the feathers (**II**). They are also homozygous dominant for an independent gene that brings about coloured plumage (**CC**). They are white because **I** is epistatic over **C**. 'White Wyandotte' poultry carry the recessive alleles of the inhibition gene (**ii**) but are white because they are also recessive for coloured feathers (**cc**). When 'White Leghorns' are mated with 'White Wyandottes' (**IICC × iicc**) the F_1 are white (**IiCc**). Interbreeding among the F_1 gives an F_2 comprising white and coloured birds in a 13 : 3 ratio. This comes about because all genotypes containing at least one dominant allele of the inhibition gene (**II**, **Ii**) are white and the double recessive (**iicc**) is also white due to lack of pigment. Only the **iiCC** and **iiCc** genotypes can be distinguished due to the presence of colour with no inhibition. If the Punnett square is drawn out showing the F_2 genotypes, the coloured birds will be seen to make up $\frac{3}{16}$ of the progeny.

Cyanogenesis in white clover

Another example of epistasis is cyanogenesis in white clover, a case involving two pairs of unlinked genes with complete dominance between alleles.

Two genes control the production of hydrogen cyanide (HCN) in white clover. The first gene has the symbol **Ac** for the dominant allele, which, when present, causes the production of cyanogenic glucosides. No glucosides are produced in plants that are homozygous for the recessive allele (**ac ac**).

The second gene is for an enzyme called linamerase, which is produced when the dominant allele (**Li**) is present, but is absent in plants homozygous for the recessive allele (**li li**). The enzyme converts cyanogenic glucosides into hydrogen cyanide.

Thus, plants with dominant alleles for both genes have the potential to produce hydrogen cyanide and are said to be strongly cyanogenic. Plants that produce glucosides but not the enzyme are mildly cyanogenic due to the slow and spontaneous hydrolysis of glucosides, which is not catalysed by enzymes. Plants without the ability to produce glucosides (that is, of genotype **ac ac**) are non-cyanogenic. This is true regardless of whether the enzyme is present, since there is nothing to be converted into hydrogen cyanide.

In a dihybrid cross between parents of genotype **Ac Ac Li Li** and **ac ac li li** all the F_1 offspring will be of genotype **Ac ac Li li** and be strongly cyanogenic, as expected. Crossing of the F_1 progeny cannot, however, give a 9 : 3 : 3 : 1 ratio among the F_2 progeny, as there only three phenotypes. Instead, a 9 : 3 : 4 ratio of strongly cyanogenic : mildly cyanogenic : non-cyanogenic phenotypes is observed, as shown in Figure 8.4, overleaf.

male gametes

	AcLi	Acli	acLi	acli
AcLi	strongly cyanogenic	strongly cyanogenic	strongly cyanogenic	strongly cyanogenic
Acli	strongly cyanogenic	mildly cyanogenic	strongly cyanogenic	mildly cyanogenic
acLi	strongly cyanogenic	strongly cyanogenic	non-cyanogenic	non-cyanogenic
acli	strongly cyanogenic	mildly cyanogenic	non-cyanogenic	non-cyanogenic

female gametes

Figure 8.4 Inheritance of cyanogenesis in white clover: the alleles present in the pollen and eggs of the F_1 progeny are shown along the top and side of the Punnett square. The phenotypes of the F_2 progeny that result are shown, demonstrating the $9:3:4$ ratio

In normal conditions, cyanogenesis is prevented from occurring in **Ac Li** plants by the enzyme and substrate being separated from one another. This is because the reaction can damage plant tissues. However, tissue damage caused by grazing animals or frost, for example, can bring the two chemicals together, causing cyanogenesis to occur. Because of this, there tend to be fewer cyanogenic genotypes in cold countries where frosts are common. However, in warmer countries, and in places where slugs and snails are a problem, there is some advantage to cyanogenesis as the animals are discouraged from grazing plants that produce hydrogen cyanide.

Summary

◆ Alleles of some genes interact with one another to give F_1 phenotypes that show incomplete dominance or co-dominance. Their heterozygotes can be distinguished from the homozygotes and this produces a modified F_2 phenotypic ratio of $1:2:1$.

◆ When two different genes control the same single character they may also interact epistatically to give rise to modifications of the familiar $9:3:3:1$ ratio.

◆ Deviations from the $3:1$ ratio also occur when one of the homozygous classes in an F_2 is determined by a lethal gene.

Questions

1 Coat colour in short-horn cattle is determined by a co-dominant pair of alleles, C^R and C^W. Cattle with genotype $C^R C^R$ have red coats while those with genotype $C^W C^W$ have a white coat. Heterozygous cattle have coats with a mixture of red and white hairs, known as 'roan'. The presence of horns is controlled by a gene with alleles **h** and **H**. The gene isn't linked to that for coat colour. Cattle of genotype **hh** have horns, those with allele **H** are hornless.

  a What is the genotype of horned roan cattle?

  b What are the expected genotype and phenotype ratios of calves born to roan cows mated with roan bulls?

  c What are the expected genotypes and phenotypes of calves born to horned roan cows mated with heterozygous hornless white bulls?

2 Hair length in guinea pigs is determined by a gene with dominant allele **L** for short hair and recessive allele **l** for long hair. Co-dominant alleles at an unlinked locus specify hair colour such that **YY** = yellow, **YW** = cream and **WW** = white. From matings between dihybrid 'short, cream' guinea pigs, **LlYW,** state the expected phenotypic ratio in the progeny.

3 Two genes control feather colour in parakeets. One has alleles **B** for the production of blue pigment and **b** for no blue pigment. The other has alleles **Y** for the production of yellow pigment and **y** for no yellow pigment. Birds homozygous recessive for only one of the genes are either yellow or blue. Those with **B** at one locus and **Y** at the other have green feathers. Some birds with white plumage also occur.

  a Explain the terms *allele, locus, genotype* and *phenotype*.

  b Make a list of all the possible genotypes for plumage colour. Beside each genotype write down what phenotype you would expect to see.

  c Explain why some parakeets have green feathers. Why do some have white feathers?

  d What phenotypic ratios would you expect in the offspring of matings between **BbYy** birds?

4 Radishes can be long **LL**, round **ll**, or oval **Ll**.

  a If equal numbers of long and oval radishes were allowed to cross freely what shapes would the offspring be? What genotypic and phenotypic ratios would you expect among the offspring?

  b How might round radishes be produced from oval ones? Why can round radishes not be produced from long ones?

  c If long radishes are crossed to round radishes and then the F_1 allowed to cross at random among themselves, what phenotypic ratio is expected in the F_2?

5 A recessive mutant allele of the gene responsible for the synthesis of chlorophyll in the tomato gives an albino plant when homozygous. Albinos die as seedlings, after using up all the food reserves in their seeds. Heterozygotes are pale coloured but survive. A normal plant was crossed with a heterozygous plant and the seeds harvested. The seeds gave rise to progeny, which were then self-pollinated. The selfed seeds were planted and the progeny found to be in the ratio of 5 normal to 2 pale-coloured plants. Explain these results.

6 A lethal gene occurs in Kerry cattle. Calves homozygous for the recessive allele are malformed and die soon after birth. Heterozygous cattle are short but viable; they are called Dexter cattle and have good beef qualities. Show diagrammatically the cross between a Dexter cow and bull. What proportion of viable calves born from such a cross would be Dexter cattle, and what proportion Kerry cattle?

7 If a sex-linked gene is a recessive lethal, causing embryo death and re-absorption at an early stage, what proportion of offspring born would be expected to be female? What proportion of offspring would be expected to carry the allele if one parent did?

8 Some cats have white patches on their coats. This effect is produced by the action of the spotting gene, **S**. This gene has two co-dominant alleles, S^1 and S^2. The coats can have large patches of white, small white patches or no patches at all.
a Explain the meaning of the term *co-dominance*. [2]
b A cat with no white patches, homozygous for S^1, was mated with a cat that had small white patches. Some of the offspring produced had coats with small white patches and the rest had no white patches. Draw out the genetic diagram to show this cross and include the expected ratio of phenotypes on your diagram. [4]

London AS/A Biology and Human Biology Module Test B/HB1 January 1999 Q5

9 Coat colour in mice is controlled by a single gene, with the dominant allele **F** representing yellow fur and the recessive allele **f** representing grey fur. The alleles are not sex-linked.
 In a large number of crosses, the coat colour of the resulting offspring was recorded. The crossing of two yellow mice produced similar results on each occasion. When the results of 30 such crosses were combined, the numbers of offspring of each type were as follows:

- yellow   208
- grey   96

a (i) Express these results in a simple ratio. [1]
(ii) State the genotypes of the male and female yellow parents in these crosses. [1]
(iii) Draw a genetic diagram to show the *expected* results of crossing these individuals. Give the expected phenotypic ratio. [3]
In all crosses involving mice, a small proportion of the offspring would not be expected to survive, being born dead. These are referred to as stillbirths. The incidence of stillborn offspring is much higher in the crosses between the yellow mice than might normally occur. If, however, yellow mice are crossed with grey mice there are very few stillbirths.
b (i) Suggest how this information can explain the difference between the phenotypic ratios given in **a (i)** and **a (iii)**. [2]
(ii) With the aid of a genetic diagram, explain the low incidence of stillborn offspring when yellow and grey mice are crossed. [4]
c Suggest why variation in coat colour in mice may be important in natural selection. [2]

UCLES A Modular Sciences: Central Concepts in Biology June 1998 Section A Q4

9 Multiple alleles and blood groups

In all Mendel's experiments with peas the individual characters that he studied were each represented by two alternative forms. The single genes controlling these characters existed as two alleles, one of which was fully dominant over the other. The other examples of monohybrid inheritance that we have so far encountered have likewise involved genes with only two alleles, although we now know that the dominance between them may be modified by various forms of interaction (Chapter 8).

There is no particular reason, of course, why a gene should have only two alleles. Genes are complex structures made up of linear sequences of hundreds of nucleotide pairs in DNA (Chapter 12) and they mutate to give a number of different allelic forms; that is, **multiple alleles**. In this chapter, we will deal with multiple alleles and also with the blood group genetics of humans. Blood groups provide the best known examples of multiple alleles and it is therefore convenient to discuss the two topics together.

What are multiple alleles?

In diploids each gene is represented twice, once on each of the two homologous chromosomes at corresponding loci. A gene with two alleles may therefore have three different genotypes – two kinds of homozygote and a heterozygote (Figure 9.1).

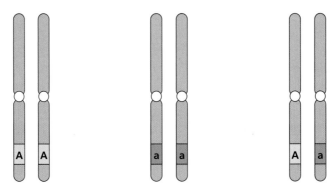

Figure 9.1

With complete dominance, there will be only two phenotypes, but in cases of incomplete dominance, and co-dominance, the heterozygote may also be distinguishable (Chapter 8).

The term 'multiple alleles' refers to the alleles of a gene for which there are more than two. Suppose a gene has three alleles – W, w^a and w^b – specifying three different forms of the same character. The gene is still represented only twice, at corresponding sites on the chromosome pair concerned, but we will now have more than three genotypes when we combine the alleles two at a time – in fact there are six possibilities (Figure 9.2).

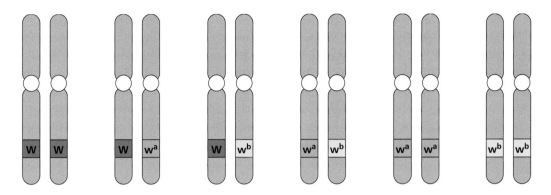

Figure 9.2

The number of phenotypes depends on which alleles are dominant over which others, and whether or not there is any incomplete dominance, or any co-dominance. With four or more alleles the situation is correspondingly more complex. In experimental matings, however, we will generally be concerned with only two parents, and only two pairs of alleles, at any one time – as in a monohybrid cross.

Multiple alleles in humans
ABO blood groups

The ABO blood group system is one of several different blood group systems in humans that are genetically determined. Phenotypes are due to proteins that act as antigens bound to the surface of the red blood cells. In the ABO system the blood groups are controlled by a single gene, designated as I, which has three alleles, I^A, I^B and I^O. Allele I^A produces antigen A, I^B produces antigen B and I^O does not produce any antigen. I^A and I^B are co-dominant and both are dominant over I^O. There are six possible pair combinations of the alleles, giving six genotypes within the human population, and four different phenotypes, as shown in Table 9.1.

Table 9.1

Genotypes	Antigens	Blood group phenotypes
$I^A I^A$	A	A
$I^A I^O$	A	A
$I^B I^B$	B	B
$I^B I^O$	B	B
$I^A I^B$	AB	AB
$I^O I^O$	–	O

Persons of blood group A carry antigen A and may be either homozygous for allele I^A (I^AI^A), or heterozygous (I^AI^O). Those of group B phenotype may likewise be homozygous (I^BI^B) or heterozygous (I^BI^O) and they carry the B antigen. People in group AB are known to be heterozygotes (I^AI^B) and to carry both the A and B antigens, while those in group O have neither A nor B antigens on their red cells and are also of known genotype (I^OI^O).

It is a simple matter to say what phenotypes and genotypes are expected in the children of parents with various blood groups (Figure 9.3).

Group O × group O

parents	I^OI^O × I^OI^O
gametes	all I^O all I^O
children	all I^OI^O

Group AB × group O
A single child has a probability of $\frac{1}{2}$ of being group A and $\frac{1}{2}$ of being group B.

parents	I^AI^B × I^OI^O
gametes	$\frac{1}{2}$ I^A $\frac{1}{2}$ I^B all I^O
children	$\frac{1}{2}$ I^AI^O $\frac{1}{2}$ I^BI^O
	group A group B

Group AB × group AB
For a single child the probability of being group A = $\frac{1}{4}$, group AB = $\frac{1}{2}$ and group B = $\frac{1}{4}$. (You could work out for yourself the outcomes of the various other kinds of crosses that can be made: A × A; B × B; A × B; A × AB; A × O; B × AB; B × O; and AB × O.)

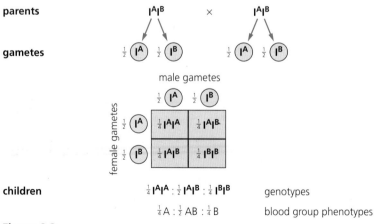

	parents	I^AI^B × I^AI^B
	gametes	$\frac{1}{2}$ I^A $\frac{1}{2}$ I^B $\frac{1}{2}$ I^A $\frac{1}{2}$ I^B

male gametes

female gametes	$\frac{1}{2}$ I^A	$\frac{1}{2}$ I^B
$\frac{1}{2}$ I^A	$\frac{1}{4}$ I^AI^A	$\frac{1}{4}$ I^AI^B
$\frac{1}{2}$ I^B	$\frac{1}{4}$ I^AI^B	$\frac{1}{4}$ I^BI^B

children	$\frac{1}{4}$ I^AI^A : $\frac{1}{2}$ I^AI^B : $\frac{1}{4}$ I^BI^B	genotypes
	$\frac{1}{4}$ A : $\frac{1}{2}$ AB : $\frac{1}{4}$ B	blood group phenotypes

Figure 9.3

An understanding of the genetics of the ABO blood group series in humans is of considerable practical importance in relation to blood typing and blood transfusion. These aspects are dealt with in Box 9.1.

Box 9.1 Blood typing and blood transfusion

Karl Landsteiner experimented, in 1901, with mixing blood from different people. He separated the red cells out by centrifugation and then mixed the cells of one person's blood with the serum of another. In some mixtures the serum caused the red cells to clump together, or agglutinate, and in others it did not. It was discovered that the clumping was caused by natural antibodies present in the serum that reacted with the *A* and *B* antigens on the surface of the red cells. Since a person cannot carry antibodies that would agglutinate his or her own cells, there are no anti-*A* or anti-*B* antibodies in the serum of people of group AB. Group O has no *A* or *B* antigens, and carries both anti-*A* and anti-*B* antibodies; and groups A and B have antibodies against each other's antigens (Figure 9.4). The agglutination reaction is shown in Figure 9.5.

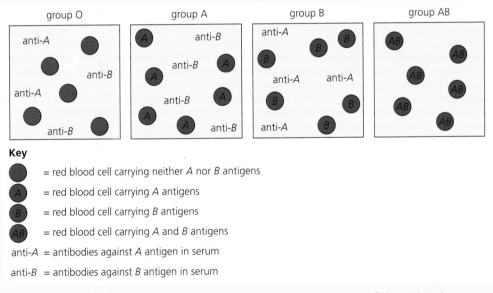

Key

- = red blood cell carrying neither *A* nor *B* antigens
- (A) = red blood cell carrying *A* antigens
- (B) = red blood cell carrying *B* antigens
- (AB) = red blood cell carrying *A* and *B* antigens

anti-*A* = antibodies against *A* antigen in serum

anti-*B* = antibodies against *B* antigen in serum

Figure 9.4 Antigens and antibodies found in the four blood group types of the ABO series

Figure 9.5 Agglutination of the red cells will occur when a drop of blood from a person of group A is mixed with serum from a person of group B. The B serum contains anti-*A* antibodies, which react with the *A* antigens on the surface of the red cells and cause them to clump together **(a)**. When the same blood is mixed with the serum of another person of blood group A there is no agglutination **(b)**

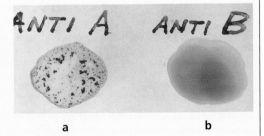

a   b

Why do some people carry natural antibodies against one another's red cell antigens? The answer, it seems, is that these antigenic substances are not confined to red blood cells of human beings: they are common to many organisms, including the bacteria that are constantly in contact with our bodies. We are therefore exposed to them in our day-to-day lives and make antibodies against them in the same way as we would with any other foreign proteins that entered our bodies. Antigens of the same type as a person has on his or her own red cells are not recognised as foreign, and no antibodies are made against them.

Blood typing

The existence of natural antibodies makes blood group classification (typing) a simple matter. Two kinds of serum are used for making tests. Serum containing anti-A antibodies is prepared from a person of blood group B, by removing the cells, and serum with anti-B antibodies is prepared from the blood of an individual of group A. A small drop of the blood to be typed is mixed with the two kinds of serum and the clumping reaction noted. A person whose blood is not agglutinated by either kind of serum is group O (that is, it has no antigens). One whose cells are agglutinated by anti-A, but not anti-B, is group A (the cells have A antigens). Conversely cells agglutinated by anti-B, but not anti-A, belong to group B (they have B antigens). An individual whose cells are clumped by both kinds of serum is in group AB (the cells have A and B antigens). The reactions are summarised in Figure 9.6.

serum type	reaction to serum of red blood cells from group:			
	O	A	B	AB
anti-A	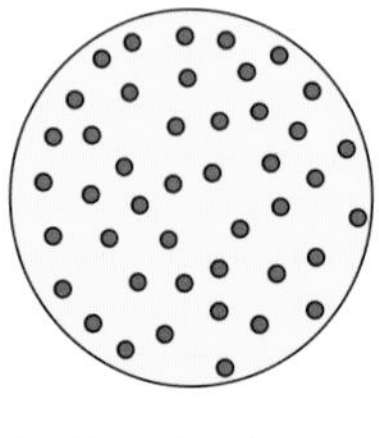	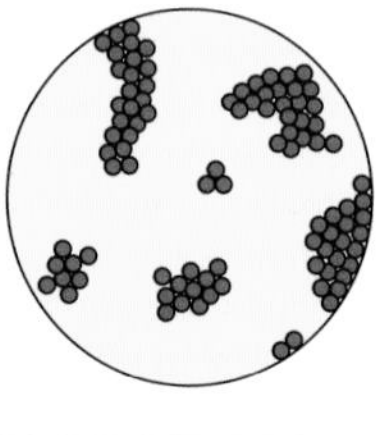	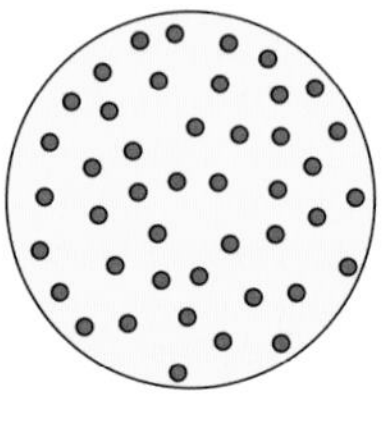	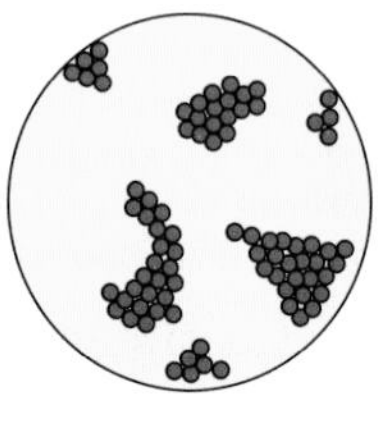
anti-B	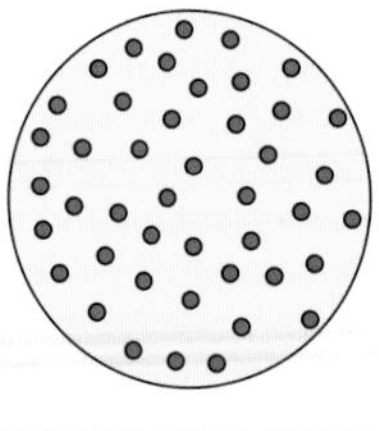	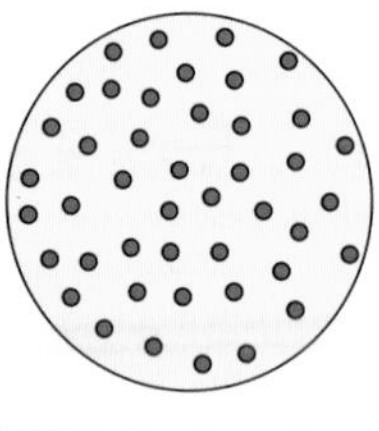	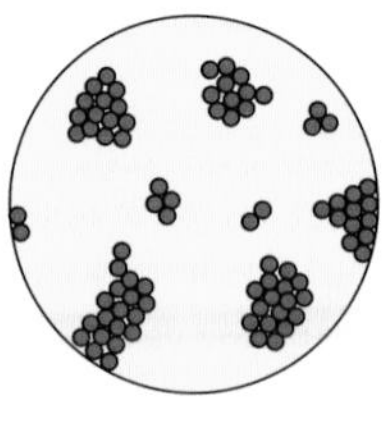	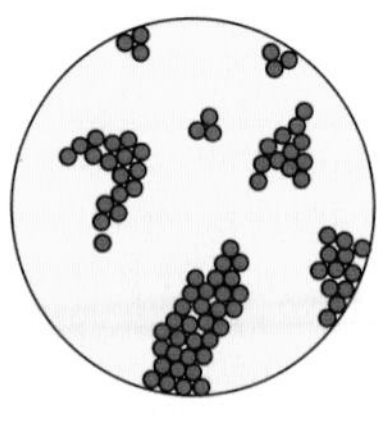

Figure 9.6 Blood groups are typed according to the way their red cells react when mixed with two kinds of serum, containing anti-A and anti-B antibodies

Blood transfusion

Blood can be *safely* given by transfusion from one person to another only when both donor and recipient are of the same group. If the groups do not match, there is a danger that the red cells of one person, or both of them, will be clumped by the antibodies of the other's serum. The reaction that matters most is that which occurs between the donor's red cells and the recipient's serum. The reason is that the donor usually contributes a small quantity of blood that is mixed with a much larger volume of the recipient's blood. If the recipient's serum contains antibodies against the donor red cells, then because of the unequal quantities the incoming red cells will be completely agglutinated. They will cause blockage of the capillaries and may result in the death of the recipient. If the donor's serum has antibodies against the recipient's red cells, the effect is much less damaging. This is because the incoming antibodies are diluted out in the large volume of the recipient's blood. They are also absorbed by various other tissues of the body and quickly removed from the circulation.

When small quantities of blood are transfused some groups can therefore be mixed with others, as explained in Table 9.2.

Table 9.2 Summary of agglutination reactions of a donor's cells, which occur when blood of different groups is mixed during transfusions. Agglutination is denoted by a plus sign and these combinations have to be avoided. The donor's antibodies are not shown because the important factor in transfusion is the way the donor's red cell antigens react with the recipient's antibodies – as explained above

Donor	Recipient O – anti-*A*, anti-*B*	A – anti-*B*	B – anti-*A*	AB
O	–	–	–	–
A (*A*)	+	–	+	–
B (*B*)	+	+	–	–
AB (*AB*)	+	+	+	–

Group O can donate to any of the others because O has no *A* or *B* antigens on its red cells and cannot be agglutinated by antibodies in the recipient's serum. For this reason people of blood group O are referred to as **universal donors**.

AB blood has no anti-*A* or anti-*B* antibodies and can receive blood from any of the other groups: it is unable to agglutinate their red cells. Persons of group AB are known as **universal recipients**.

Blood group A cannot be given to people of groups O or B, since the anti-*A* antibodies in their serum would clump the incoming cells immediately. Similarly, B type blood is not donated to groups O or A.

MN blood groups

The MN blood groups are another case of multiple alleles in humans. They are determined by a different gene from that of the ABO system. In introductory texts it is usual to confine the discussion of them to two main alleles and to make only passing reference to the fact that there are many subdivisions of alleles within the main ones – as indeed there are in the ABO system.

As mentioned in Chapter 8, the two alleles are co-dominant. Each one produces an antigen on the red cells so that homozygotes (**MM, NN**) have either the M or N antigen and the heterozygotes (**MN**) have both. There are thus three different genotypes in the population and three blood group phenotypes (Table 9.3).

Table 9.3

Genotype	Antigen	Phenotype
MM	*M*	M
NN	*N*	N
MN	*MN*	MN

Inheritance of the gene is straightforward. The six different crosses that can be made are given below together with their outcome in the progeny (Table 9.4).

Table 9.4

Parents	Children	Parents	Children
1 **MM** × **MM**	All **MM**	4 **MM** × **MN**	$\frac{1}{2}$ **MM** : $\frac{1}{2}$ **MN**
2 **NN** × **NN**	All **NN**	5 **NN** × **MN**	$\frac{1}{2}$ **NN** : $\frac{1}{2}$ **MN**
3 **MM** × **NN**	All **MN**	6 **MN** × **MN**	$\frac{1}{4}$ **MM** : $\frac{1}{2}$ **MN** : $\frac{1}{4}$ **NN**

There are no natural antibodies in serum against the M and N antigens – and therefore no problems with transfusions. To classify MN blood types it is necessary to prepare two kinds of antiserum (anti-M and anti-N) by injecting human red cells into rabbits to induce antibody formation. Rabbit blood containing the antibodies is then used to obtain purified antiserum, which can be used in agglutination tests.

Rhesus blood groups

Rhesus blood groups were described by Landsteiner and Wiener in 1940 when they found that antiserum prepared by injecting blood of the rhesus monkey into rabbits and guinea pigs could agglutinate the red cells of humans. Of the people tested, 85% were shown to carry an antigen that reacted with the rhesus antibodies, and 15% did not. Those with the antigen are called 'rhesus-positive' types and those without are 'rhesus-negative'. The difference was later shown to be due to a single gene with two alleles, symbolised as **Rh$^+$** and **Rh$^-$**. **Rh$^+$** is fully dominant over **Rh$^-$** so that heterozygotes have the rhesus-positive phenotype (Table 9.5).

Table 9.5

Genotypes	Antigen	Phenotypes
Rh^+Rh^+	+	rhesus-positive
Rh^+Rh^-	+	rhesus-positive
Rh^-Rh^-	−	rhesus-negative

Table 9.6

Parents	Children
$Rh^+Rh^+ \times Rh^+Rh^+$	All Rh^+Rh^+
$Rh^+Rh^+ \times Rh^+Rh^-$	$\frac{1}{2} Rh^+Rh^+ : \frac{1}{2} Rh^+Rh^-$
$Rh^+Rh^- \times Rh^+Rh^-$	$\frac{1}{4} Rh^+Rh^+ : \frac{1}{2} Rh^+Rh^- : \frac{1}{4} Rh^-Rh^-$

Inheritance of the gene follows the typical Mendelian pattern for a single gene with two alleles. Rhesus-negative parents have only rhesus-negative children. The outcome of matings between two rhesus-positives depends upon whether the parents are both homozygous, both heterozygous or one of each (Table 9.6).

It is now known that the alleles of the **Rh** gene form a multiple allelic series and that suitable immunological tests can distinguish many subdivisions of the two main alleles – but we will not concern ourselves with these details.

One of the most important aspects of the 'rhesus' antigens is their involvement in the blood disorder **erythroblastosis fetalis**, which occurs when the rhesus blood types of the mother and her fetus are incompatible. The details are explained in Box 9.2.

Box 9.2 Erythroblastosis fetalis

The condition of erythroblastosis fetalis arises when a rhesus-negative mother (Rh^-Rh^-) carries a fetus that has a Rh^+ allele transmitted by the father. The fetus is then heterozygous (Rh^+Rh^-) and produces rhesus antigens on its red blood cells; that is, it has the rhesus-positive phenotype. Leakage of blood across the placenta, if some damage occurs at or near to birth, can allow rhesus-positive cells from the fetus' blood into the mother's circulation – and she then builds up antibodies against them in her blood serum. The mother thus becomes 'sensitised', and when she carries a second rhesus-positive child her antibodies may be present in high enough concentration to agglutinate (clump) and break down the red cells of the fetus. Such antibodies can pass freely across the placenta from mother to fetus. The problem is not as great as may be thought, however, since the father is rhesus-positive and the mother rhesus-negative in only about one in eight sexual partnerships; and then only one in 200 infants are affected by the disease. It is also now possible to give protection against the condition by injecting the mother with anti-Rh serum immediately after the birth of the first rhesus-positive child. This treatment destroys any rhesus-positive cells that may have entered her blood while she was carrying the child, and prevents the build up of antibodies in her system.

For reasons that we cannot go into here, this kind of incompatibility between mother and fetus is far less troublesome for other blood group systems.

Legal aspects of blood group genetics

Blood groups are useful characters for resolving legal cases about such problems as baby 'mix-ups' in maternity hospitals, disputed paternity cases and forensic science. The main reason is that the characters are so distinctive, easily classified and determined by single genes with simple patterns of inheritance. Their importance in relation to disease and blood transfusion also means that extensive records exist of the blood types of a large number of the population, whereas this kind of information would not necessarily be available for other phenotypic characters. Another factor, which is of importance in forensic science, is that the blood group of a person can be determined from traces of blood left behind at the scene of a crime long after the perpetrator has gone. In the case of the ABO blood groups, it is also possible to classify the blood type without using an individual's blood at all. Some people are 'secretors' and produce the A and B antigens in certain other body fluids such as saliva, tears and semen.

The main point about using blood groups as evidence is that they can never prove that an individual was involved in a particular misdemeanour, but they can definitely *exclude* him or her. If a woman claimed, for instance, that a certain man was the father of her child, and she was blood group O, the baby was O and the accused was AB, then clearly he would be innocent of that charge. If, however, his blood group was O this would not prove that he *was* the father – because there are numerous other males in the population with that blood group.

Example

A farmer has two sons. The first, born when the farmer was young, grew into a handsome healthy youth, in whom he took great pride. The second, born much later, was always a sickly child and neighbours' talk induced the farmer to bring his wife to court, disputing the child's paternity. The grounds for the dispute were that the farmer, having produced so fit a first son, could not be the father of a weakling.

The blood groups were as shown in Table 9.7.

Table 9.7

	ABO	MN
farmer	O	M
mother	AB	N
first son	A	N
second son	B	MN

What advice can we give the court?

If we look at the ABO system, the farmer was O and therefore $I^O I^O$ in genotype. The mother was AB and therefore $I^A I^B$ in genotype. The cross was therefore as shown in Figure 9.7.

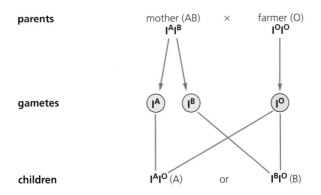

Figure 9.7

We can see that these parents can have two sons who have A or B blood types, and on this basis both of the sons could have come from the marriage.

When we consider the MN system, however, we see that the farmer must have been **MM** (as he was phenotype M) and the mother must have been **NN** (as she was phenotype N) so the children can only be **MN** (Figure 9.8).

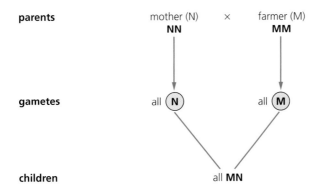

Figure 9.8

Since the first son was N blood type (and therefore **NN** genotype) we would have to tell the farmer that he could have fathered the second son but that he was definitely not the father of the first.

Other blood group systems

In addition to the three genes discussed above there are at least twelve others in humans that determine lesser known, and less important, blood group systems, such as Kell, Lutheran, Duffy, Kidd, Lewis and so on. They are all controlled by different independently inherited genes with two or more alleles. These blood group genes have their own specific antigen types, all of which are bound to the surface of the same red blood cells. When considered all together, they make it possible to describe a really detailed 'genetic profile' of the blood group phenotype of an individual person.

Summary

◆ Genes may have more than two alleles.

◆ As a result of mutation they can have several different forms – that is, multiple alleles – which determine several different phenotypes for a single character.

◆ The best known example is the gene controlling the ABO blood group system in humans, which has three main alleles.

◆ When a gene is represented by multiple alleles, only two of them may be present at the locus concerned in any one individual.

Questions

1 A family tree is shown below with the blood group phenotypes for some individuals.

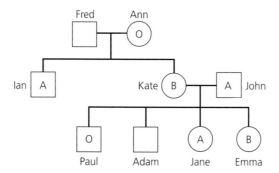

a Using the symbols I^A, I^B and I^O to represent the alleles, write down the genotype of each person whose phenotype is shown.

b What possible genotypes could the other two people have?

c What are the dominance relations between the three alleles involved in determining this blood group?

2 During ten years of marriage Mrs X failed to become pregnant. She then had an affair with Mr Y and during the next five years had three children. From the blood types shown below what can you determine about the paternity of each child?

Person	ABO	MN
Mrs X	O	MN
Mr X	O	MN
Mr Y	A	N
child 1	O	MN
child 2	O	M
child 3	A	N

3 A mother in a maternity hospital claimed that she had not been given her own baby. There was only one other baby in the ward at the time. The husband of the mother is now dead but they have three other children. From the following information work out which baby belongs to the mother.

Person	ABO	MN	Ability to taste phenylthiocarbamide
mother	O	MN	'non-taster'
baby a	A	M	'taster'
baby b	O	M	'non-taster'
child 1	A	MN	'taster'
child 2	B	N	'taster'
child 3	A	MN	'non-taster'

4 The inheritance of ABO blood groups is controlled by three alleles of the same gene, I^A, I^B and I^O. The alleles I^A and I^B are co-dominant. Both I^A and I^B are dominant to the allele I^O.

a Explain what is meant by an *allele*. [1]

b (i) Copy and complete the table to show the missing genotypes. [2]

Blood group phenotype	Possible genotypes
A	$I^A I^A$, ___
B	$I^B I^B$, ___
AB	___
O	___

(ii) Children of blood groups A and O were born to parents of blood groups A and B. Copy and complete the genetic diagram to show the possible ABO blood group

parental phenotypes	blood group A	×	blood group B
parental genotypes	___		___
genotypes of gametes	___		___
genotypes of children	___		
phenotypes of children	___		

phenotypes of the children which could be produced from these parents. [3]

AQA (NEAB) AS/A Biology: Continuity of Life (BY02) Module Test March 1999
Section A Q2

Continuous variation

We have so far confined our studies to characters controlled by one or two genes that have distinctive and clear-cut alternative forms. The alleles of the genes concerned have major effects on the phenotype, which are readily distinguished, and cannot be confused with any additional variation that may be due to the environment. This kind of clear-cut difference in the forms of various characters is known as **discontinuous variation** (for example, 'tall' and 'short' peas).

It is essential to begin the study of genetics in this way in order to identify individual genes and to explain the basic principles of the subject in the clearest possible terms. At the same time, it is important to emphasise that in nature characters of the visible phenotype that show only a few discontinuous and easily classified forms are the *exception* rather than the *rule*. Most of the natural variation that we see about us, and which is thought to be the most important in terms of natural selection and evolution, and in agriculture, is not of this discontinuous kind at all. If we look at the more obvious variations within species, such as their size, weight and morphology, we see that these characters seldom fall into categories that are easily classified, or grouped, into any simple pattern. On the contrary, they are represented by numerous forms, and though there may be a range of sizes, from small to large, the intermediates differ from each other by very little, and so form a range that is known as **continuous variation**. Characters of this kind are much more difficult for the geneticist to work with. They are usually determined by several genes, each of which has a small effect, and which are not readily identified as individual units of heredity.

In this chapter, we briefly explain how we can study the inheritance and the genetic basis of continuous variation, and see why it is that we normally give so little attention to it in our elementary studies. At the end of this chapter, we also summarise everything that we have so far encountered about the relation between genes and characters.

Continuous and discontinuous variation

It is helpful at the outset to explain precisely what we mean by the terms 'continuous' and 'discontinuous' variation. We will deal with discontinuous variation first because this is the pattern of variation with which we are already familiar. If we examine a population of pea plants of mixed heights, say from the backcross **Tt** × **tt**, we will find the situation depicted in the histogram in Figure 10.1a, overleaf. The plants fall into two discrete and discontinuous classes without any intermediate forms. They are either 'tall' or 'dwarf' (see Figure 3.4, page 33). We saw the same clear-cut pattern of variation in the pea seed-colour phenotypes, of 'yellow' and 'green', and in the white, pink and red flower colours of the four-o'clock plant (Chapter 8). Many other similar examples can be listed – yellow and purple seeds in maize, horned and polled (hornless) cattle, vestigial and normal wings in *Drosophila*, brown and blue eye colour in humans, and so on.

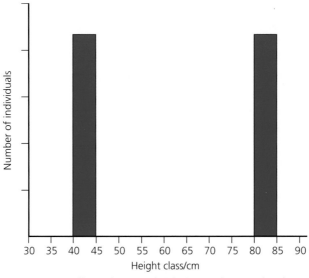

a discontinuous – height in peas (mature plants)

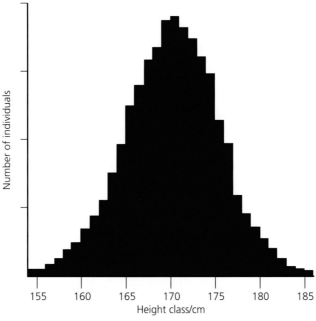

b continuous – height in humans (adult males)

Figure 10.1 Frequency histograms showing the difference between continuous and discontinuous patterns of variation. 'Height' in peas is a character that shows discontinuous variation **(a)**. Plants fall into two discrete and quite separate groups, and individuals can easily be classified as either 'tall' or 'dwarf'. The character 'height' in humans shows a more continuous variation **(b)**. There are numerous forms that are not so discrete or separate, but which blend into one another giving a more continuous range or distribution. If the height classes are divided into intervals that are small enough, the pattern of distribution becomes virtually a smooth curve

Characters that fall into a few easily classifiable types in this way, without a range of intermediates, are said to display **qualitative variation**. The different forms are recognised by their *qualities*, such as 'tallness' and 'dwarfness', and we don't have to make any measurements in order to classify them. There may be some minor variations between one tall pea plant and another, and also among the individual short ones, which are due to environment; but because the difference in the two character forms is determined by alleles of a single major gene, with large effects, we have no difficulty in distinguishing between them.

Continuous or **quantitative variation**, on the other hand, means that the character has numerous forms that are not discrete, or separate, but which all merge into one another to give a continuous distribution of values. 'Height' in humans is an example. In a population such as that shown in the frequency histogram in Figure 10.1b there are no discrete types to be found with respect to this character. We cannot place individuals into only two or three categories of 'tall', 'short' or 'intermediate height', because they do not fall into such a simple pattern of variation. There is a continuous distribution of heights spread throughout the range 150–190 cm.

The bulk of the population is grouped around the mean height value of 170 cm with the rest tailing off on either side in a symmetrical pattern. If enough measurements are taken they can be divided into so many height classes that the histogram becomes a smooth line curve. A symmetrical frequency histogram of this kind is known as a **normal curve**. The population can be described in terms of the **mean** and the **standard deviation** (that is, the spread about the mean).

To describe any one individual in the population shown in Figure 10.1b we would have to measure the height and classify the person in terms of cm and mm. In other words, the character is quantitative and individuals have to be 'quantified' (measured, in this case) in order to describe their phenotype. The vast majority of characters of living organisms are of this quantitative kind and show a pattern of continuous variation that fits a normal curve. Another well-known example is the character of 'intelligence' in humans. Others include the weight and size of most plants and animals, the yield of crop plants, the egg-laying capacity of hens, milk yield in cows, and so on.

It is difficult to deal with the genetics of quantitative characters, simply because the variations do not fall into a few simple classes, as they do with qualitative variations, and we cannot identify the individual genes that control them (unless they are very few in number). Generally speaking, characters that show continuous variation are controlled by a large number of different genes, each of which has only a small effect. Their individual contributions are therefore 'lost' against the background of environmental variation, and it becomes difficult in most cases to sort out the influence of environment from that of genotype.

Johannsen's pure line experiment

How do we know that quantitative characters are controlled by genes? The Danish geneticist Johannsen gave us the answer to this question with his detailed breeding experiments on the dwarf bean, *Phaseolus vulgaris*, carried out in 1903.

The character that Johannsen studied was 'seed weight'. He chose to work with the dwarf bean because it is a self-fertilising plant and all the descendants of one seed are what he called a 'pure line'; that is, all of the seeds taken off one plant are genetically identical. Pure lines are homozygous at all their gene loci and they breed true. Different lines are homozygous for different combinations of dominant and recessive alleles; for example, **AAbbccDDee** ... or **aaBBCCddEE** ...

Johannsen began his experiments with a collection of 19 pure lines obtained from different sources. By using lines of diverse origin he knew that they would be genetically distinct. Each line had a characteristic mean seed weight. It ranged from 35.1 centigrams (cg) in line 19, to 64.2 cg in line 1. Within the lines there was variation in seed weight due to environmental factors, such as the position that a seed occupied within the pod and the position of the pod on the plant.

In one experiment, Johannsen mixed together seeds from all 19 lines. A frequency histogram of their weights gave a continuous distribution over the range 5–95 cg, with an overall mean of 48 cg. Although the mean weights of the lines all differed from one another, the variation within lines, due to environmental effects, was so large that it 'masked' these mean differences and so gave rise to the continuous pattern of variation. He then demonstrated that it was possible to change the mean seed weight in progeny grown from samples of this mixture by *selection*. He picked out two samples, one of small seeds and the other of large seeds. These were grown and the progeny seeds they produced were weighed. The progeny differed in their mean weights. Those coming from the 'small parents' were lighter than those coming from the 'large parents' (Figure 10.2a).

When the same selection procedure was repeated using the variation *within lines* there was no response (Figure 10.2b). Johannsen concluded from this that variation in the character of seed weight could be separated into two components: *heritable* and *non-heritable* variation. Seeds from different lines vary as a result of genetic and environmental factors. Selection on the mixture works because in the process of picking out large and small seeds different genotypes are being taken, and it is the heritable part of the variation that is being transmitted to the progeny to give the difference in mean seed weight between the two samples. Seeds from the same pure line have no genetic differences. All their variation is due to the environment and is non-heritable. Progeny from samples of large and small seeds within lines have identical mean weights because their parents have the same genotype.

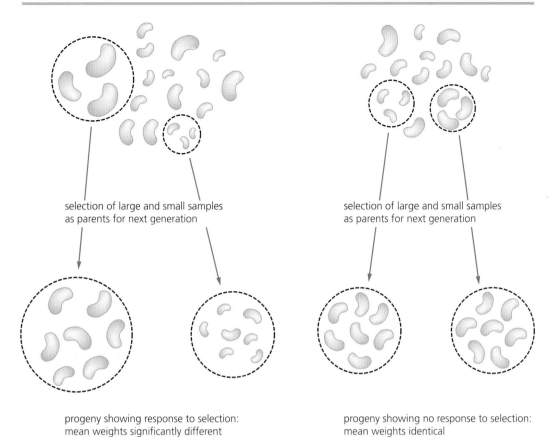

selection of large and small samples
as parents for next generation

selection of large and small samples
as parents for next generation

progeny showing response to selection:
mean weights significantly different

progeny showing no response to selection:
mean weights identical

a mixture of lines: genetic + environmental variation

b pure line: environmental variation only

Figure 10.2 Johannsen's experiment with the dwarf bean showed that quantitative characters are controlled by genes, and that with inbred pure lines it is possible to distinguish between heritable and non-heritable variation

The importance of Johannsen's work was in showing that for a quantitative character part of the variation at least is controlled by genes, and that by using the appropriate experimental procedure it is possible to 'dissect' the variation into its two component parts and to distinguish between what he called the *genotype* and the *phenotype*. The genotypic component is that which is represented by the mean seed weights of the lines, and the phenotypic part is the variation about the mean within lines.

In outcrossing species (for example, cross-pollinating species of plants), there are no pure lines and every individual is a different genotype (Chapter 6). In these cases it is obviously much more difficult to unravel the two components of continuous variation.

Johannsen showed quite convincingly that quantitative variation was due to genes as well as to environment. But he wasn't able to relate the differences between his lines to any *particular* genes, or to identify any individual units of heredity.

Multiple factors

Following Johannsen's work, there was controversy among geneticists as to whether continuous variation could be accounted for on the basis of Mendel's idea of discrete units of inheritance. Some favoured the suggestion that the blending of phenotypes could easily be explained if the character was controlled by several different genes, each with a small cumulative effect. Others argued that Mendel's unit factors were not the answer and that some other, entirely different, type of hereditary factor was involved.

The issue was resolved in about 1910 when firm evidence for the **multiple factor**, or **multiple gene**, hypothesis was provided by the Swedish geneticist Nilsson-Ehle, and subsequently by several others. Nilsson-Ehle worked with grain colour in wheat (*Triticum aestivum*). In one of his experiments, he crossed a pure-breeding variety with dark red grains to one with white grains (wheat is also a self-fertilising species). The F_1 had grains of intermediate colour. In the F_2 there was a ratio of 15 coloured : 1 white, but the coloured grains varied in the density of their red pigmentation. The character was not *qualitative* because the F_2 could not be described as simply red or white. It was *quantitative* and each grain had to be classified on a scale of five shades of colour from white through to dark red. As the shades were not easily distinguished, it could be argued that the variation was practically continuous. By selfing the F_2 plants, and looking at their breeding behaviour in the F_3, Nilsson-Ehle was able to confirm their genotypes and to show that the character was controlled by only two genes (Figure 10.3).

The genes concerned behaved in their transmission like any other Mendelian genes. The only difference in this instance was that the two genes worked in such a way that their effects were *cumulative* or *additive*, and each dominant allele contributed a certain degree of 'redness' to the colour expression of the phenotype. In the model shown in Figure 10.3, we have assigned an arbitrary value of one unit to the colour contribution of each dominant allele and a value of zero to the recessives. On the basis of **additive gene action** it is now a simple matter to see how the independent segregation of only two pairs of alleles from heterozygous F_1 can give a range of five phenotypic classes in the F_2. If an element of environmental variation is included as well it becomes even easier to appreciate how a character controlled in this way, by relatively few genes, can approximate to a continuous distribution.

It is important to understand that in this experiment we are dealing with two genes determining *one* character, not two separate characters as in Mendel's work, and we do not expect to find a 9 : 3 : 3 : 1 ratio in the F_2. Here the genes are acting in an *additive* manner and we get an effect on the phenotype that is directly proportional to the number of dominant alleles present in the genotype.

In practice, of course, most cases of continuous variation are not so amenable to study as this one, and in most cases we have no idea how many different genes are involved and no simple means of finding out. The value of Nilsson-Ehle's experiment, and others like it, lay in emphasising how continuous variation could be understood in terms of simple Mendelian genetics. There is no need to invoke anything more than multiple genes to explain it. The word **polygenes** is now used in preference to 'multiple genes' to describe the genes that control a quantitative character. The term **polygenic inheritance** was introduced by Mather to describe characters whose expression is controlled by many genes with individual slight effects on the phenotype. The main features of polygenic inheritance are summarised as follows.

1 The characters are controlled by a number of genes.
2 The genes have individual small effects and changing one allele for another one at a locus causes relatively little difference in the phenotype.
3 The phenotype is subject to considerable environmental variation.
4 Characters show a continuous range of variation.

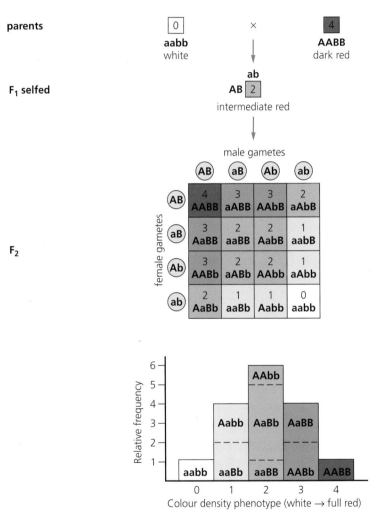

Figure 10.3 Nilsson-Ehle's experiment on the genetic basis of grain colour in wheat. A cross between two pure-breeding strains that differ by two genes with additive (cumulative) effects gives an almost continuous range of variation in the F_2. Each dominant allele contributes one unit of colour density to the character

Genes and characters

At this stage, it may be helpful to collect together our ideas on the relation between genes and characters, since it is obviously not as simple as we suggested in some of the earlier chapters. In Figure 10.4, it is assumed that the inheritance of the genes is according to Mendelian principles of segregation and independent segregation, and that what we are concerned with here are the different forms of action and interaction of genes, at the level of the visible phenotype, rather than any complications in their transmission.

Genes and characters

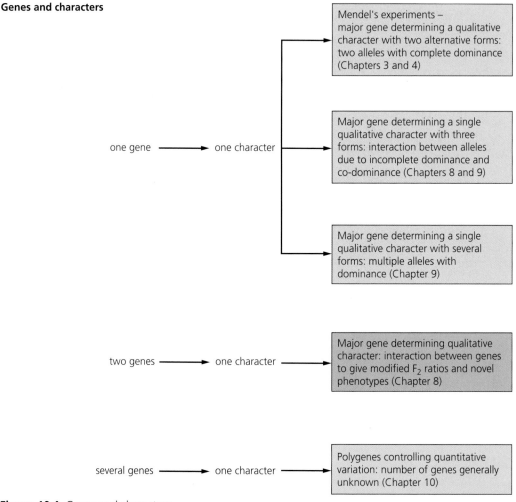

Figure 10.4 Genes and characters

Figure 10.4 is not meant to be taken as a complete list of all possible modes of gene action, but only to put together those aspects that we have covered so far. We will have more to say about gene action in later chapters when we come to deal with the utilisation of genetic information at the molecular level (Chapters 12 and 13).

Summary

◆ Characters that show continuous variation are controlled by several genes (polygenes).

◆ The individual contribution of the genes to the phenotype is small, relative to that of the environment, and it is difficult to identify them as separate units of inheritance.

◆ Where this can be done the experiments suggest that the number of polygenes is not necessarily high and that they are transmitted according to the same laws as major genes.

Further reading

1 W. S. Clug and M. R. Cummings (2000) *Concepts of Genetics*. Sixth edition. Prentice Hall, Inc. Website:
http://cw.prenhall.com/bookbind/pubbooks/klug3/
2 *MendelWeb*. Resource for teachers and students interested in classical genetics, introductory data analysis, and elementary plant science. Includes texts, illustrations, bibliographies and links to further resources. Mendel's original paper is given in German, with a revised version of the English translation:
http://www.netspace.org/MendelWeb

Questions

1 The chart below illustrates the typical distribution of the heights of sweet pea plants.

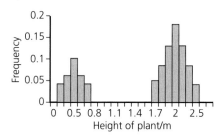

Use your knowledge of the causes of genetic variation to explain this pattern of variation.

2 Explain, with examples, what is meant by *continuous* and *discontinuous variation*. How does such variation arise?

3 A cross was made between two varieties of pure-breeding maize, one with short ears, called Tom Thumb, and one with long ears, called Black Mexican. The lengths of ears from plants in the F_1 and F_2 generations were measured (to the nearest centimetre). The frequencies of different ear lengths in the parents and progeny are shown in the histograms below.

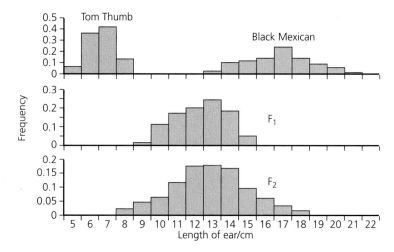

a The amount of variation for the F_1 is between that shown for the parents. That for the F_2 is greater than for either parent or the F_1. Explain this observation.

b What term is used to describe the type of variation displayed by the character 'ear length'?

c What is the genetic explanation for this type of variation?

4 Distinguish between each of the following pairs of terms, giving examples where appropriate.
  a *Autosomal linkage* and *sex linkage*   [3]
  b *Continuous variation* and *discontinuous variation*.   [3]
                  London AS/A Biology and Human Biology Module Test B/HB1 June 1998 Q4

DNA and chromosomes

We have covered the basic principles of genetics and discussed how the inheritance of genes is related to the behaviour of chromosomes during division of the nucleus. We have also discussed some aspects of the way in which genes act, and interact with one another, in determining characters. In all of this we have regarded genes, as Mendel did, simply as 'units of heredity'. Until the 1950s this was the only view that geneticists could have, because before that time little was known about the structure and composition of the genetic material. The laws, or principles, of heredity were all worked out without any firm knowledge of the nature of the genes or of the chromosomes that carry them. In the 1950s, the chemical nature of heredity was discovered. The identification of the molecular structure of the genetic material led to a new understanding of the gene.

We now know that the genetic material is made up of a chemical substance called deoxyribonucleic acid (DNA). We will deal with the molecular structure and replication of DNA molecules and show how they are organised within the chromosomes. In Chapter 12, we explain how the information in DNA is utilised and what precisely we mean by 'the gene', in molecular terms.

1 **Chromosomes contain DNA:** Chemical analysis shows that chromosomes contain DNA. There is a complete parallel between the way in which genes and chromosomes are distributed during nuclear division (Chapter 5). This parallel is embodied in the chromosome theory of heredity and strongly suggests that the chromosomes carry the genetic information.

2 **Constancy of DNA content of nuclei:** Diploid nuclei from somatic cells in any one species, at corresponding stages of the mitotic cycle, all contain the same quantity of DNA. Gametic nuclei have half the amount, as would be expected from genetic experiments.

3 **Stability of DNA:** In contrast to other cell constituents, DNA remains stable and intact as a large macromolecule. It is not metabolised.

Although the evidence that DNA is the genetic material became established beyond doubt in 1944, one problem still puzzled biologists. How could such a relatively simple molecule fulfil all the properties of the genetic material? The answer was found in the *structure* of DNA.

The structure of DNA

Although the structure of DNA itself was unknown before the 1950s, the constituent building blocks had earlier been isolated and their structure determined from chemical analysis of purified DNA molecules.

Constituents of DNA

DNA can be broken down into three constituents: a pentose (or 5-carbon) sugar called deoxyribose (Figure 11.1), organic bases and phosphoric acid.

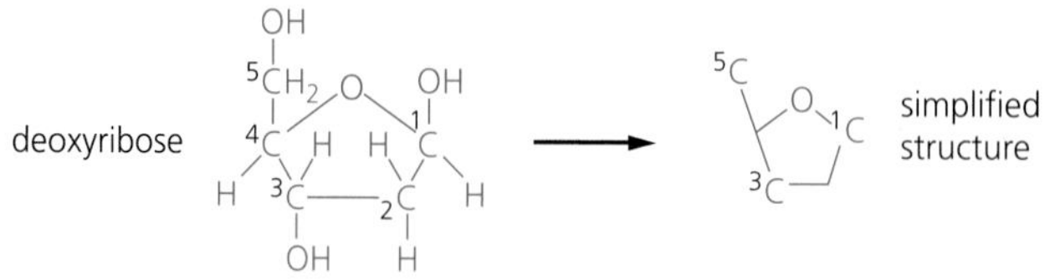

Figure 11.1

The organic bases are of four types: adenine (**A**), guanine (**G**), thymine (**T**) and cytosine (**C**). They are ring compounds, which include carbon and nitrogen molecules. They fall into two groups – purines and pyrimidines – depending upon whether the ring is a double or single structure. Phosphoric acid can be denoted as Ⓟ; its structural formula, and those of the organic bases, is given in Figure 11.2.

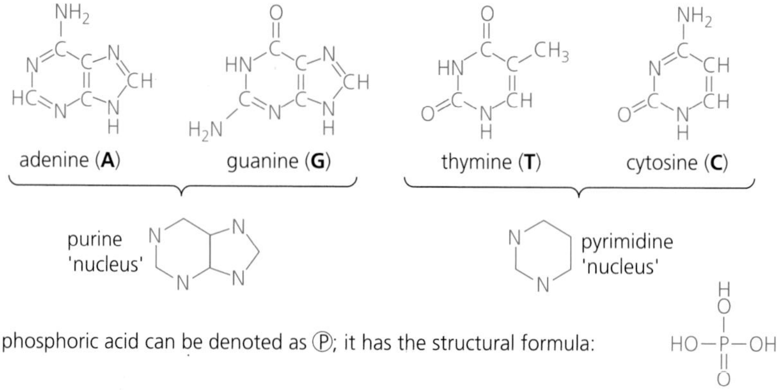

phosphoric acid can be denoted as Ⓟ; it has the structural formula:

Figure 11.2

The nucleotide

The basic unit structure of the DNA molecule is the **nucleotide**. This contains one of each of the three kinds of constituent molecule – a base, a sugar and a phosphate group (Figure 11.3a). The base is attached to the sugar at carbon-1 (^1C). The phosphate group is linked to the ^5C atom. The nucleotide can be shown as a simple diagram (Figure 11.3b) in which shapes are used to represent the constituent parts. There are four different nucleotides corresponding to the four different bases: all of them have identical sugar and phosphate molecules. By a process of **polymerisation**, nucleotides link together to form long chains called **polynucleotides**.

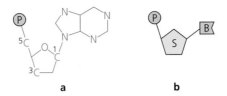

Figure 11.3 Diagrams showing the structure of a nucleotide (S = sugar, P = phosphate, B = base)

The polynucleotide

In the polymerisation process, the linking of nucleotides is brought about by condensation reactions in which the phosphate groups bond alternately from position ^{5}C of one sugar molecule onto position ^{3}C of the next one, so forming a long ...–sugar–phosphate–sugar–phosphate–... 'backbone'. When a polynucleotide chain 'grows', the nucleotides are added to the hydroxyl (OH) group at the ^{3}C position of the deoxyribose sugar. In other words, they are always added at only one end of the chain, which is known as the 3' (3-prime) end. The other end of the chain is the 5' end. Because of the way in which the chains are formed, they have a *direction*, with 'growth' at the 3' end (Figure 11.4).

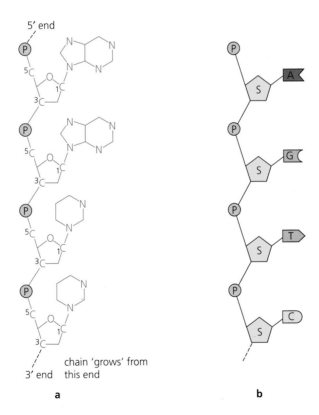

Figure 11.4

The double helix

Once it became known for certain that DNA was the genetic material, and knowledge began to accumulate about the chemical nature of its constituent parts, many biologists were keen to unravel the secrets of its three-dimensional structure.

In 1953, James Watson and Francis Crick, working in the Cavendish Laboratory in Cambridge, were the first to solve this puzzle. They showed how the polynucleotides were organised within the DNA molecule. A personal account of the events leading up to their discoveries is given in Watson's book *The Double Helix*. In this book, Watson refers to the major contributions made by other scientists – notably Maurice Wilkins and Rosalind Franklin of King's College London. In working out their model, Watson and Crick drew together knowledge from two main lines of enquiry:

1 chemical analysis
2 X-ray crystallography.

Chemical analysis

In 1949, Erwin Chargaff had already made an important discovery about the base composition of the DNA extracted from a number of different species. Chargaff found that irrespective of its source the DNA always showed equivalent amounts of the purine base adenine (A) and the pyrimidine base thymine (T) – ($A:T = 1:1$). The same was true for guanine (G) and cytosine (C). In contrast there was no such fixed relation between the quantities of the two purines, $A:G$, or the two pyrimidines, $T:C$. Nor was there any relation between $A:C$ and $G:T$. These findings, which became known as **Chargaff's rules**, can be summarised as follows:

1 The number of purine bases ($A + G$) = the number of pyrimidine bases ($T + C$).
2 The number of adenine bases = the number of thymine bases ($A:T = 1:1$).
3 The number of guanine bases = the number of cytosine bases ($G:C = 1:1$).

Obviously these rules imposed some restrictions on the way in which the bases could be arranged within the DNA molecule. Watson and Crick's interpretation of Chargaff's rules was that the base A was always paired with T, and G with C. This could only be achieved if DNA consisted of two strands held together by specific base-pairing.

X-ray crystallography

Wilkins and Franklin had been working on X-ray diffraction patterns for some time. When X-rays are passed through crystalline preparations of DNA, they are scattered in a certain way according to the arrangement of atoms within the molecules. The scatter pattern can be recorded by allowing the X-rays to impinge on a photographic film. Spots on the film reveal information about the angle of scatter and the underlying three-dimensional structure of the molecules in the crystal (Figure 11.5).

Identical X-ray diffraction patterns were obtained for DNA taken from T2 bacteriophage, various bacteria, trout and bulls. These pictures showed that DNA was a long thin molecule with a constant diameter of 2.0 nm, and that it was coiled in the form of a helix. The helix made one full twist for every 3.4 nm of its length. As there was a distance of only 0.34 nm between each pair of bases there were reckoned to be ten bases per twist. The density of atoms in the molecule also indicated that it must be composed of *two* strands.

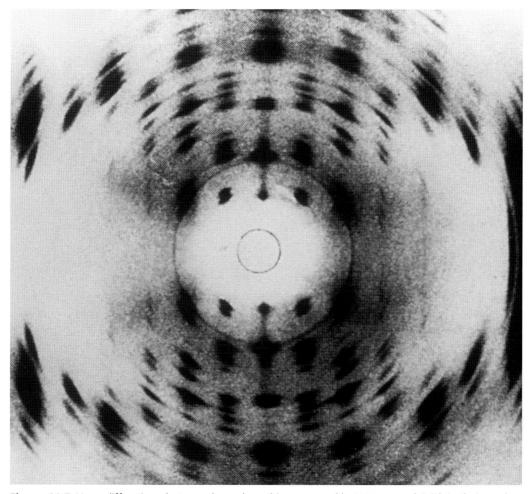

Figure 11.5 X-ray diffraction photographs such as this were used by Watson and Crick in their work on the structure of DNA. The photographs were taken by Rosalind Franklin in Maurice Wilkins' laboratory. The helical structure of the molecule is indicated by the repeated pattern of dark areas surrounding the centre

Having assembled this information, and taken into account certain other relevant facts about the angles of bonds and the distances between atoms, Watson and Crick set about building a model – using pieces of wire and flat metal shapes to represent bases. They came to the conclusion that the only way to fit all the constituents together was in the form of a *double* helix composed of two polynucleotide chains held together by hydrogen bonding between pairs of bases (Figure 11.6, overleaf).

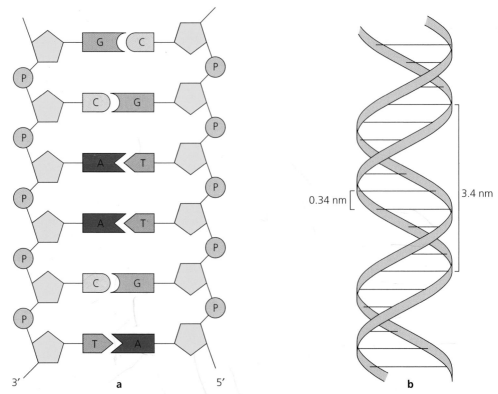

Figure 11.6 The DNA molecule is made up of two complementary polynucleotide strands that are anti-parallel (that is, they run in opposite directions) **(a)**. Hydrogen bonding between specific pairs of bases holds the two strands together. The strands are twisted into a double helix **(b)**. Continuous 'ribbons' represent the sugar–phosphate backbone and horizontal lines represent the base-pairs

The salient features of the model may be summarised as follows.

1 The DNA molecule is a *double helix* made up of two polynucleotide chains that are coiled around the same axis and interlocked. They can only be separated by untwisting – not simply pulling apart sideways.

2 The sugar–phosphate backbones are on the outside of the molecule and the bases are on the inside.

3 The chains are held together by *hydrogen bonding* between specific pairs of bases, according to Chargaff's rules. Adenine only pairs with thymine (**A**:**T**) and guanine only pairs with cytosine (**G**:**C**). These particular combinations give the most efficient 'lock and key' arrangement of hydrogen bonding and are the only pairs that can fit together within the dimensions of the molecule (Figure 11.7).

4 Because of the specific base-pairing, the sequence of nucleotides in one strand determines the sequence in the other one. The two strands are therefore *complementary*.

5 The bases in the two strands can only be made to fit together if the sugar molecules to which they are attached point in opposite directions; that is, the strands are *anti-parallel*.

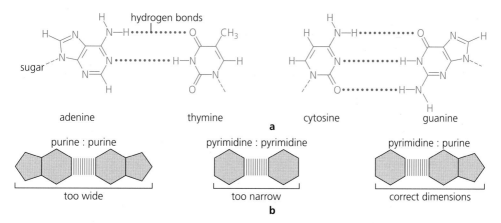

Figure 11.7 The two polynucleotide strands of DNA are held together by hydrogen bonds that form between specific pairs of bases **(a)**. Pyrimidine:pyrimidine and purine:purine pairs of bases have the wrong dimensions to fit within the double helix **(b)**

The molecule can be thought of as a ladder in which the base-pairs are the rungs and the sugar–phosphate backbones represent the two sides. The ladder is then twisted into a double helix in which there are ten nucleotides per turn. The bases are flat structures at right angles to the long axis of the molecule.

In 1962, Watson and Crick, together with Maurice Wilkins, were awarded the Nobel Prize for Medicine for their work on elucidating the structure of DNA. Rosalind Franklin had died in 1958, and so could not be awarded the prize.

Size of DNA molecules

In Figure 11.6, only a tiny stretch of double helix is shown. In reality the molecules are extremely long and can be composed of several million nucleotides. It is known that each chromatid of a eukaryotic chromosome contains one continuous molecule of DNA double helix running throughout its length, as explained later. The size of the chromosome depends on the length of the DNA molecule from which it is (mainly) comprised.

Implications of the Watson and Crick model of DNA structure

There are two important features of the Watson and Crick model in terms of the arrangement of bases.

1 There are no restrictions on the order in which the nucleotides can occur along one of the polynucleotide strands, and this suggests that the *sequence* of bases may be important as a means of storing and encoding the genetic information. This is precisely what Watson and Crick proposed. We deal with this aspect in Chapter 12.
2 **Complementary base-pairing** means that for any given sequence of the four bases in one of the strands, the sequence in the other one is determined. This suggested a mechanism by which DNA could self-replicate and make more identical copies of itself. The way in which this replication works is explained overleaf.

Replication of DNA

When Watson and Crick published an account of their model for DNA structure they also put forward a theory for the way in which it could undergo **replication**. According to this idea, the two strands could unwind from one another, following the disruption of their hydrogen bonds, and each one could then serve as a *template* for the synthesis of a new complementary strand. The molecule could 'unzip' its bonds from one end, and then new bases present in the nucleus could be assembled alongside their complementary partners, in the correct sequence, and become linked up by the formation of a new sugar–phosphate backbone.

When the unwinding and synthesis had passed along the full length of the molecule, two identical daughter molecules would result, each of which would be an exact copy of the original double helix. This was later called **semi-conservative replication**, because of the way in which the newly formed daughter molecules each contained (or 'conserved') one of the original 'parental' strands, plus one newly synthesised strand. The mechanism is illustrated in Figure 11.8. Later on, this model was shown to be correct.

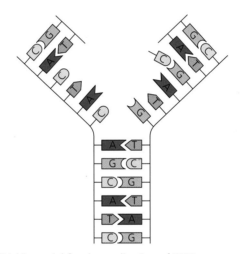

Figure 11.8 Watson and Crick's model for the replication of DNA

From the bacterium *E. coli*, Kornberg (1957) purified the enzyme DNA-polymerase, which is responsible for the 'zipping up' of bases during the synthesis of new polynucleotide chains. He was then able to use this enzyme to direct the synthesis of new molecules of DNA in a test tube. Single-stranded DNA templates, of known base composition, were mixed together with the polymerase enzyme and a supply of free nucleotides. Double-stranded DNA was then formed in which the base sequence of the new strand was complementary to that of the existing polynucleotide. This experiment showed that the replication of DNA was indeed based on specific pairing between complementary bases. (However, it didn't prove that the process was *semi-conservative*; this was demonstrated later in an experiment by Meselson and Stahl in 1958.)

DNA and chromosome structure

The chemical composition and molecular structure of DNA are normally studied using highly purified samples extracted from cells. In its *natural* form, of course, DNA is not a highly purified crystalline substance. It is a component of the chromosomes. We have already discussed chromosomes (Chapter 2) in relation to the distribution of genetic information during division of the nucleus in mitosis and meiosis. We now look into the molecular organisation of DNA in the chromosomes.

There are in fact two kinds of chromosomes: the simple naked ones found in viruses, bacteria and cell organelles, and the more complex ones that occur in the nuclei in eukaryotes. We will be concerned principally with the latter.

Prokaryotic chromosomes

The chromosomes of prokaryotes are simply long naked molecules of double-stranded DNA. The bacterium *E. coli*, for example, has a single circular chromosome, about 1.4 mm in length, which contains four million nucleotide pairs. It replicates in a semi-conservative fashion, beginning at a fixed point and proceeding in one direction around the length of the chromosome at a rate of 16 000 base-pairs per minute. Relatively simple chromosomes of this type are also found in many DNA viruses, as well as in chloroplasts and mitochondria of higher organisms.

Eukaryotic chromosomes

In eukaryotes the nucleus of each cell contains several pairs of chromosomes. The chromosomes are large structures and they change their form and structural organisation at different stages throughout the cell cycle (Chapter 2). They are composed of about equal amounts of protein and DNA. The DNA, of course, carries the genetic information and the proteins are mainly concerned with structure. The chemical complex of DNA and protein that is found in chromosomes is known as **nucleoprotein** or **chromatin**. The proteins are of two types – basic **histones** and acidic proteins (non-histones). There is a large variety of non-histones, but only five types of basic histones (H1, H2A, H2B, H3 and H4), which are common to most species.

Eukaryotic chromosomes contain an enormous quantity of DNA. In *Drosophila*, for example, the largest chromosome of the complement has 62 000 000 nucleotide base-pairs in each chromatid. The largest one in humans has more than 200 000 000.

The sheer quantity of DNA in a eukaryotic chromosome poses a problem of organisation. The organisation must allow for the changes that take place during cell division. At interphase, the DNA needs to extend itself in order to replicate and allow for the activity of genes, while in nuclear division it must become tightly packed and shortened in order for the chromosomes to move and to divide during mitosis and meiosis. Consequently, the DNA is not present in a fully extended form in the chromosome. It is 'packaged' in some way so that 8.5 cm of DNA helix that is found in chromosome 1 of humans, for instance, is all accommodated within a metaphase chromosome that is only 10 μm long. In this case, the packing ratio of extended DNA : metaphase chromosome, is almost 10 000 : 1. At interphase, when the chromosomes are much less tightly coiled, the packing ratio is about 100 : 1.

Box 11.1 Organisation of DNA in chromosomes

When chromatin from an interphase nucleus is spread out and viewed under the electron microscope, it has a definite repeating structure that resembles a string of beads. The 'beads' are **nucleosomes**. They consist of an 'octamer' of two molecules each of histones H2A, H2B, H3 and H4, around which is wrapped one and a third turns of DNA (Figure 11.9a). There are 146 nucleotide pairs within the DNA encircling the nucleosome, and when the length of double helix that runs to the next nucleosome is included, the total length is about 200 nucleotides. The nucleosome is the basic unit of structure of chromatin. The nucleosome fibre shown in Figure 11.9b is what we see when interphase chromatin is spread out for analysis; in its natural state, the chromatin fibre (solenoid fibre) is three times thicker (Figure 11.9c and d). Packing of nucleosomes into the solenoid fibre is thought to be the main function of histone H1, which is attached to the outside of the nucleosome.

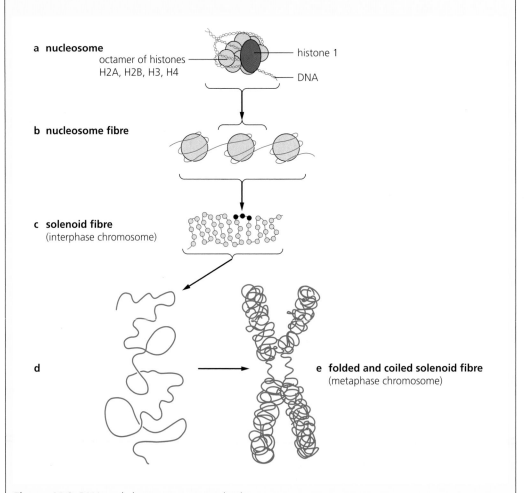

a nucleosome
octamer of histones
H2A, H2B, H3, H4
histone 1
DNA

b nucleosome fibre

c solenoid fibre
(interphase chromosome)

d

e folded and coiled solenoid fibre
(metaphase chromosome)

Figure 11.9 DNA and chromosome organisation

The turns that the DNA makes about the nucleosome give a packing ratio of 10:1, and with further coiling into the solenoid fibre a ratio of 100:1 is achieved. When the chromatin is condensed (by coiling and folding) during cell division, into a metaphase chromosome, a further packing occurs to bring the ratio up to 10000:1. Details of this final level of organisation are uncertain. By studying metaphase chromosomes treated to deplete them of histones, it can be seen that DNA loops are attached to a protein scaffold running the length of each chromatid. Thus the acidic (non-histone) proteins are thought to form the scaffold around which chromatin fibres are looped. Chromatin fibres are visible in the photograph of a metaphase chromosome shown in Figure 11.10b.

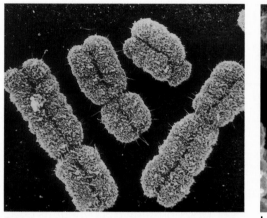
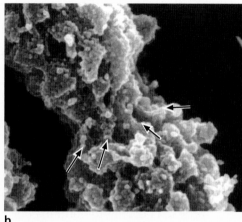

a
b

Figure 11.10 SEMs showing human chromosomes at metaphase of mitosis, at lower **(a)** and higher **(b)** magnifications. The chromatin, or solenoid, fibres (arrowed) can be seen particularly well at the higher magnification

Replication of DNA in chromosomes

Since the chromatid contains one double helix of DNA, it is able to replicate semi-conservatively, in the same way that naked DNA molecules do, during the synthesis (S) phase of the mitotic cycle. Because there is so much DNA in a chromosome, however, the process is rather complicated, as explained in Box 11.2.

At each site, the replication forks moves out in two directions so that eventually they all meet up with one another. When replication begins at a site, it is necessary for one of the DNA strands to break, in order for the two strands to unwind and allow synthesis of the new strands by complementary base-pairing. There are enzymes that can cause single-stranded 'nicks' in DNA and others that can join them back again (such enzymes are also needed for the breakage and rejoining of DNA strands that takes place during crossing over at meiosis). During DNA replication, in the interphase chromosome the existing nucleosomes remain associated with one of the newly formed chromatids. New nucleosomes are then made to organise the nucleoprotein complex of the other chromatid.

Box 11.2 Replication of DNA in chromosomes

DNA molecules in eukaryote chromosomes are too long to replicate their whole length from only one initiation point – as they do in prokaryotes. It is impossible to untwist all of the DNA in a chromatid from one end through to the other. In *Drosophila* the replication fork moves at a rate of 2600 bases per minute. At this rate it would take more than 16 days for the longest chromosome to duplicate itself if it had only one origin of replication. In fact it replicates in less than three minutes, by opening up more than 6000 replication forks at different sites along the chromosome. A short stretch of DNA with two replication sites is shown below (Figure 11.11).

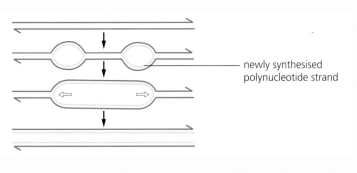

newly synthesised polynucleotide strand

Figure 11.11

DNA, chromosome structure and replication

In essence, each chromatid contains one helix of DNA running throughout its entire length. Replication of the chromatid is semi-conservative and follows the same pattern as the DNA that it carries. We can therefore represent the mitotic cycle of chromosome duplication and division in terms of the DNA helix, as given in Figure 11.12. This is the simplest possible relation between DNA and chromosomes, and overlooks the complexities of organisation discussed above.

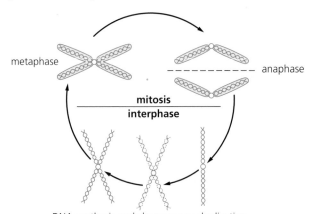

DNA synthesis and chromosome duplication

Figure 11.12 Simplified diagram showing the way in which DNA is organised in the structure and replication of the chromosome

Summary

◆ The genetic material that carries the information for heredity is a chemical substance called deoxyribonucleic acid (DNA).

◆ DNA is a long molecule made up of two chains of polynucleotides, which are twisted together into a double helix.

◆ The polynucleotides are composed of four kinds of organic bases – adenine, thymine, cytosine and guanine – which are linked together in a long chain by their attachment to sugar–phosphate backbones.

◆ Hydrogen bonding between specific pairs of bases holds the polynucleotides together and also provides the mechanism for their self-replication.

◆ The sequence of bases along a polynucleotide encodes the genetic information.

◆ In prokaryotes, viruses and cell organelles, the chromosomes are simply naked molecules of DNA. Eukaryote chromosomes are composed of nucleoprotein fibres that are a complex of DNA and protein.

◆ The organisation of eukaryote chromosomes is complicated but in essence each chromatid consists of one molecule of double helix running through its length.

Further reading

1 J. D. Watson (1970) *The Double Helix*. Penguin Books.
2 J. D. Watson and F. H. C. Crick (1953) 'A structure for deoxyribose nucleic acid', *Nature* **171**, 737–738.
3 A. D. Bates and A. Maxwell (1993) *DNA topology: In focus*. IRL Press.
4 *DNA Learning Centre*. Includes a variety of instructional materials and information for both students and teachers: **http://vector.cshl.org**
5 A. Kornberg and T. A. Baker (1992) *DNA Replication*. Second edition. Freeman.
6 A. Travers, (1993) *DNA–Protein Interactions*. Chapman & Hall.

Questions follow Chapter 12.

DNA and genes

In this chapter, we discuss the way in which genetic information is encoded in DNA. We also describe the way that it is utilised within cells and what precisely we mean by the idea of a 'gene'. First we establish the link between genes and proteins.

Genes and proteins

The idea about how genes could act was put forward as early as 1909 by the English physician Sir Archibald Garrod. At that time, the gene was a new concept, and nobody had the faintest notion about its chemical composition or structure. Garrod thought that genes might exert their effects through enzymes. He suggested that a rare human disease called alkaptonuria was due to a block in the metabolism of the amino acid tyrosine.

The disease is inherited as a simple Mendelian recessive character, and Garrod surmised that the 'inborn error of metabolism' came about because of the failure of the recessive allele (in homozygotes) to control the synthesis of an enzyme needed for the metabolism of tyrosine. The disease, although harmless, is distinctive. Because the enzyme homogentisate oxidase is lacking, homogentisic acid accumulates and is excreted in the urine, which makes it turn black after standing.

The idea that genes work through their control over the production of enzymes was therefore born very early in the history of genetics. It came to the forefront of our thinking, however, in 1941 when Beadle and Tatum published the results of their pioneering work on the biochemical genetics of the bread mould *Neurospora crassa*. They showed beyond doubt that a mutation in a single gene resulted in a change in the activity of a single enzyme.

One gene, one enzyme

Neurospora is an ascomycetous fungus that is haploid for most of its life cycle. Recessive mutations are therefore expressed in the phenotype – and readily detected. Crosses between strains are easily made and the sexual cycle is completed in as little as ten days, resulting in fruiting bodies (perithecia) full of sac-like asci. Each ascus contains eight large haploid ascospores that can be dissected out and then grown in order to test their genotype.

Normal-phenotype *Neurospora* can be cultured on minimal agar medium. The fungus has a full complement of enzymes that enables it to use simple ingredients to manufacture the essential substances needed for normal growth and development. Mutants of the normal strain can be obtained that are unable to make certain substances required for their nutrition, such as a particular amino acid or vitamin. These are known as **auxotrophic** mutants, in contrast to the normal **prototrophic** strain. Auxotrophs will only grow on minimal medium when it is supplemented with the particular substance that they are unable to make for themselves.

Beadle and Tatum isolated auxotrophic mutants of *Neurospora*, in order to use the simple biochemical defects as *characters*. The procedure they used is illustrated in

Figure 12.1. Pure cultures of the mutants were studied biochemically. Mutants accumulated certain substances normally found in only small amounts. This would happen if the pathway was blocked and the substance made in the step before the block was accumulating. It was thus argued that the defects were due to the lack of a single enzyme needed to complete the relevant biochemical pathway. The character that distinguished each mutant from the normal strain was therefore a simple biochemical one, involving the absence or presence of a single enzyme involved in the nutrition of the fungus.

When crosses were made between each of the mutants and the normal strain the character difference was inherited as a single gene with two alleles. Ascospores taken out of hybrid asci gave four normal cultures and four mutants when grown and tested. In other words, a mutation in a single gene caused the absence, or loss of activity, of a specific enzyme.

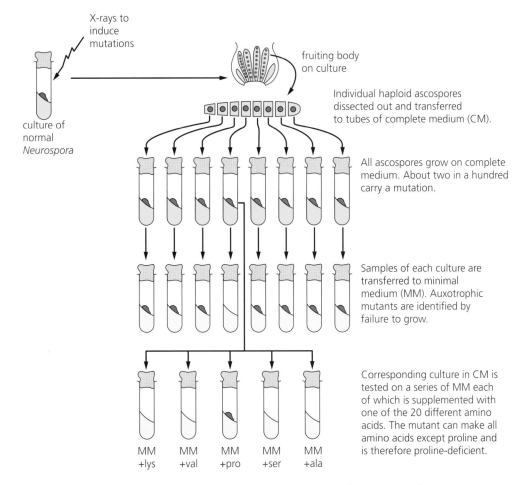

Figure 12.1 Summary of the procedure used for isolating and classifying auxotrophic mutants in *Neurospora*. The scheme shows how a mutant that is unable to make a particular amino acid is detected

These findings led Beadle and Tatum to formulate the hypothesis of 'one gene, one enzyme'. They envisaged that all biochemical processes are under genetic control; that these processes proceed through a series of stepwise reactions; and that each step is catalysed by a single enzyme and *each* enzyme is controlled by *one* gene. This vision has proved to be essentially correct, except for some minor modifications described below. Figure 12.2 shows a small part of a biochemical pathway in humans: each specific enzyme failure is associated with a well-known human genetic disorder that is inherited as a simple recessive character. It illustrates how genes direct the biochemical reactions of cells, through their control over enzymes, and thereby determine the characters of all living organisms.

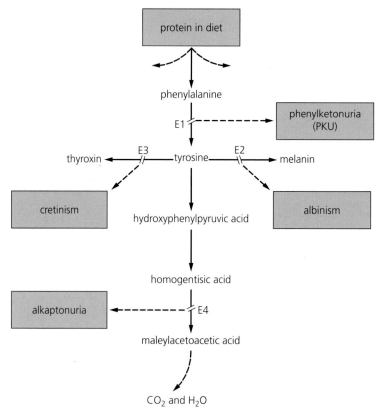

Figure 12.2 Small part of a metabolic pathway in humans. Each block in the pathway is due to a defect in an enzyme (E)

In order to understand how genes determine enzymes, we need to know something about enzymes.

Enzymes belong to a class of substances called proteins. They catalyse the biochemical reactions that take place within cells. Each enzyme has a precise structure that is necessary for its function. It seems a reasonable hypothesis, therefore, that genes might act by determining the *structure* of enzymes. Before we can examine this hypothesis we need to describe the structure of enzymes; that is, the structure of proteins.

Protein structure

A **protein** is an organic compound composed of amino acids linked together by peptide bonds to form long chains, called polypeptides. There are 20 different **amino acid** sub-units commonly found in proteins. They can all be represented by the general formula shown in Figure 12.3.

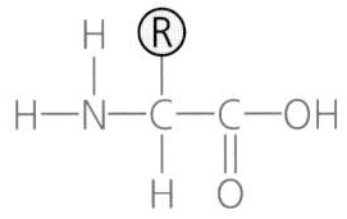

$$H-N-\overset{\overset{\displaystyle H}{|}}{\underset{\underset{\displaystyle H}{|}}{C}}-\overset{\overset{\displaystyle ®}{|}}{\underset{\underset{\displaystyle O}{||}}{C}}-OH$$

Figure 12.3

All amino acids have a carboxyl group (COOH) and an amino group (NH_2) attached to the same carbon atom. They differ only in the side-chain represented as ®. Amino acids link together by a condensation reaction between their carboxyl and amino groups, to form a peptide bond, with molecules of water being released in the process. The long chains of peptides that result from this reaction are called **polypeptides**. Each has a free amino group at one end (N-terminus) and a free carboxyl group at the other end (C-terminus).

As there is virtually an infinite number of ways in which the 20 amino acids can be assembled to make up a polypeptide, there are a limitless number of proteins that may result. Each one has a *specific sequence* essential to its structure and function. The complete amino acid sequences of several hundred different proteins are now known.

The specific sequence of amino acids in a polypeptide chain gives the protein molecule its **primary structure**. Amino acids that are close to one another in the chain then interact in various ways. This may cause, for example, one part of the linear polypeptide to coil up into an α-helix, and another part to form a β-pleated sheet. These interactions give the **secondary structure**. Disulphide bonds between sulphur-containing amino acids that are far apart in the chain stabilise the molecule in its characteristic three-dimensional or **tertiary structure**. Proteins that contain more than one polypeptide have a further level of organisation known as the **quaternary structure**.

Because these higher levels of organisation are determined by the positions that various amino acids occupy in the primary chain, the correct sequence is vital if the protein is to perform its proper function. This is especially important in the case of enzymes that have an **active site**. The loss, or incorrect order, of amino acids will lead to a corresponding loss, or reduction, in enzyme activity.

The 'one gene, one enzyme' hypothesis might therefore be explained on the basis that genes determine the amino acid sequence of enzymes.

One gene, one polypeptide chain

As we have explained, enzymes are only one class of proteins. There are many others that also play important roles in structure and metabolism.

Structural proteins such as collagen (in cartilage), keratin (in hair and hooves) and actin and myosin (in muscles) are vital components of the structure of animal bodies. *Antibody* proteins play a vital role in the body's defence mechanism against disease.

Certain *hormones* (for example, insulin) are also proteins: they take part in regulating the physiology and development of organisms. Haemoglobin is a *transport* protein involved in carrying oxygen in the blood. Histones and tubulin are important proteins concerned, respectively, with the structure of chromosomes, and with the spindle apparatus on which the chromosomes move during mitosis and meiosis.

It soon became clear that the structure of all of these proteins, and not just the enzymes, is specified by the information encoded in the genes of the DNA. It became necessary, therefore, to make the 'one gene, one enzyme' hypothesis more general, and to rephrase it as 'one gene, one *protein*'. When it was further realised that the quaternary structure of proteins may be due to two (or more) different polypeptides, and that these could be determined by quite separate genes, it was necessary to modify the hypothesis again, and to phrase the modern statement of Beadle and Tatum's principle as 'one gene, one *polypeptide chain*'. These modifications came about largely as a result of detailed studies of the structure of haemoglobin.

Haemoglobin is a transport protein that carries oxygen. It is found in the erythrocytes (red cells) of the blood and gives them their red colour. In normal adults the molecule is composed of four polypeptides – two identical alpha (α) chains and two identical beta (β) chains. These four chains associate together to give the quaternary structure of the active protein (Figure 12.4). The α-polypeptides each have an identical sequence of 141 amino acids and the β-polypeptides have 146. The two polypeptides are specified by two different genes that are unlinked.

An abnormal form of haemoglobin in some people causes them to suffer from the blood disorder called **sickle-cell anaemia**. The disease is genetically determined and inherited as a simple Mendelian character. It is expressed in recessive homozygotes and results in a malformation of the erythrocytes, which take on a 'sickle' shape (see Figure 16.4, page 210). In 1956, Vernon Ingram used a new 'fingerprinting' technique for protein analysis, and showed that the difference between normal haemoglobin (HbA) and the abnormal form (Hbs) was due to a difference of only one amino acid. Valine had been substituted for glutamic acid at position 6 near the N-terminal end of the chains (Figure 12.4).

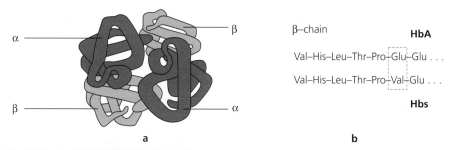

Figure 12.4 The two α and two β polypeptides that comprise the quaternary structure of haemoglobin **(a)**. The single amino acid substitution in the chain that changes HbA (normal) into Hbs (sickle-cell) **(b)**

Ingram's research clinched the idea that genes bring about their effects on characters by controlling the structure of proteins. One gene provides the information for one polypeptide chain. Mutations alter the phenotype because they affect the amino acid sequence of the proteins and change the way that the proteins carry out their function within the cell.

How can this information be carried in the DNA?

The genetic code

Once it was established that genes worked by specifying the amino acid sequence of polypeptide chains, it was obvious that their information must be carried in the sequence of bases in the DNA. The way in which the genetic information is encoded in DNA is referred to as the **genetic code**.

On purely theoretical grounds, there are various ways in which the genetic code could be organised. We can rule out the idea that one base codes for one amino acid, because we have only four kinds of bases and there are 20 different amino acids. If bases are taken two at a time, along the sequence, then there are 16 (that is 4^2) possible combinations. This takes the order into account; for example, **AT** and **TA** are different. This is still insufficient to meet the requirements.

When bases are taken three at a time, however, as triplets, we see that there are 64 (4^3) different ways in which they can be combined together. This gives us more than enough bits of information to code for 20 amino acids. In fact, we have 44 triplets to spare.

Based on such theoretical arguments, the 'base triplet hypothesis' was put forward and the triplets became known as **codons**. It would be inefficient, of course, to have a code based on any more than the minimum number of three bases per codon. Experimental evidence in support of the base triplet hypothesis was provided by Francis Crick in the early 1960s. The details of his work are beyond the scope of this text, but some reference to the kind of experiments that he carried out is given in Chapter 13.

The question we now consider is how the coded information contained in the DNA is utilised in the cell to determine the structure of proteins.

Utilisation of genetic information

In eukaryotic organisms, the DNA is located in the nucleus, while the assembly of amino acids into proteins takes place in the cytoplasm, outside of the nucleus. Information therefore has to be transferred from the site where it is stored to the site where it is utilised within the cell. Both the transfer and the utilisation are accomplished through the agency of another kind of nucleic acid known as ribonucleic acid, or simply RNA. RNA molecules are found in large quantities in cells and tissues that are actively engaged in the manufacture of proteins.

Ribonucleic acid (RNA)

RNA is a nucleic acid and is therefore similar in both its chemical composition and its structure to DNA. It is composed of nucleotides that are polymerised into chains of polynucleotides. The main features that distinguish RNA from DNA are as follows.

1 RNA is a single-stranded molecule, but it may be folded into various complex forms in which some double-stranded regions are found.
2 The sugar molecules in RNA are ribose, rather than deoxyribose as in DNA. In ribose sugar, a hydroxyl group (OH) replaces the hydrogen that is present at the ^{2}C position in deoxyribose (Figure 12.5).

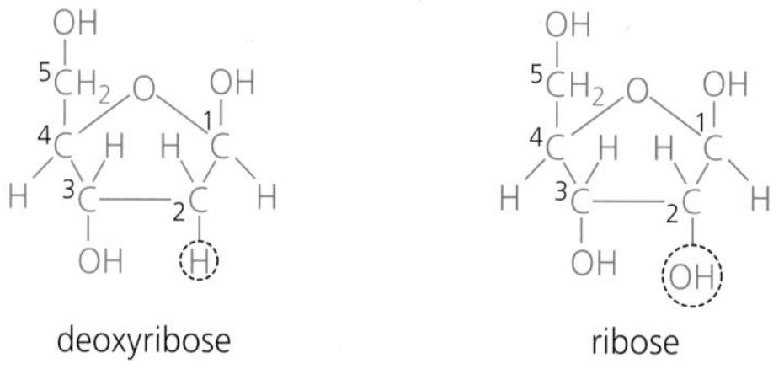

deoxyribose   ribose

Figure 12.5

3 In RNA, the base uracil (**U**) replaces the thymine in DNA (but no one really knows why).

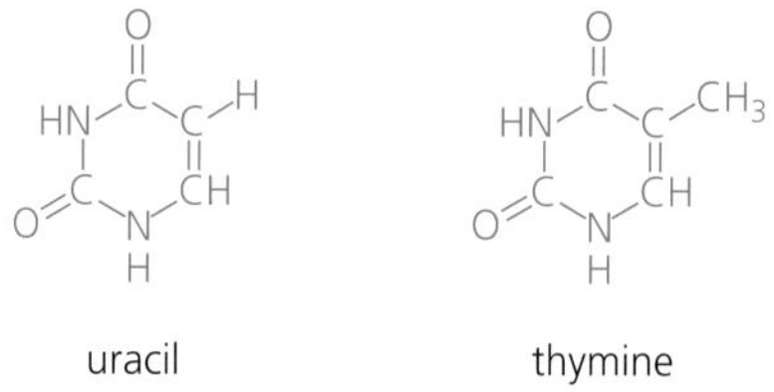

uracil   thymine

Figure 12.6

There are three different types of RNA molecules present in the cell, each of which plays a key role in the biosynthesis of proteins: these are messenger RNA, transfer RNA and ribosomal RNA.

Messenger RNA (mRNA), as its name implies, is the RNA that carries the genetic message between the DNA in the nucleus and the site of protein synthesis in the cytoplasm. In order for the message to be read off the DNA, the double helix first of all unwinds in the region of the gene, or genes, that are being expressed, and one of the strands then serves as a template for the synthesis of a complementary strand of mRNA. The process in which mRNA is synthesised from a DNA template is known as **transcription** (copying). Transcription is catalysed by a key enzyme called **RNA polymerase**. As the synthesis of the mRNA polynucleotide proceeds, it is unzipped from the DNA template, and when it is completed the mRNA is transported across the nuclear membrane and into the cytoplasm. Figure 12.7 is a diagram of transcription.

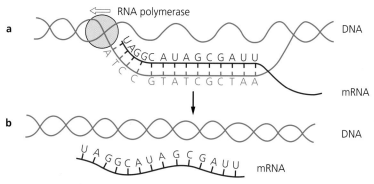

Figure 12.7 Transcription of mRNA from DNA

Transfer RNA (tRNA) is found in the cytoplasm. Its function is to pick up amino acids and to carry them to the sites of protein synthesis, on the ribosomes, so that they can be joined together into polypeptides. There are at least 20 different tRNA molecules, one for each amino acid, and they are all similar in their basic structure. They consist of a single strand of RNA, about 80 bases long, which is folded back upon itself in a clover-leaf arrangement due to pairing between complementary bases (Figure 12.8). The tRNA molecules also contain some unusual nucleotides, such as pseudouridine and inosine, which are unable to form hydrogen bonds with other bases; they are found in the unpaired loops within the molecule.

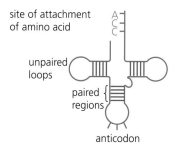

Figure 12.8 Simplified diagram of the structure of transfer RNA (tRNA). Each tRNA molecule has its own anticodon and a site of attachment for a specific amino acid

One unpaired end of each tRNA contains a triplet of exposed nucleotides, known as the **anticodon**, which is complementary to one (or more) of the codons carried in mRNA. The other unpaired end has a site for attachment to a specific amino acid. Each tRNA therefore picks up its own amino acid, and by matching of its anticodon with the complementary codon in mRNA the amino acids can be assembled in the correct sequence – as explained below.

Ribosomal RNA (rRNA) is a component of the **ribosomes**. These are the structures within the cell where the synthesis of proteins takes place. The ribosomes are often associated with the endoplasmic reticulum. They are uniform spherical structures composed of two parts – the small and the large sub-units (Figure 12.9). Each sub-unit is made up of about equal parts of RNA and of protein. The small sub-unit has one RNA molecule of approximately 1500 bases in length, and the large sub-unit has two RNAs – one of 3000 bases and one of 100 bases. The ribosomes of prokaryotes are a little smaller than those of eukaryotes, and in both groups there are several thousand of them present in each cell.

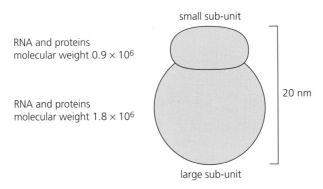

RNA and proteins
molecular weight 0.9×10^6

RNA and proteins
molecular weight 1.8×10^6

small sub-unit

large sub-unit

20 nm

Figure 12.9 Diagrammatic representation of a ribosome

Protein synthesis

Protein synthesis is the linking of amino acids to form polypeptides. It takes place on the ribosomes. The function of the ribosome is to attach to the mRNA, by its small sub-unit, and to hold the messenger in such a way that its codons can be recognised and paired with the complementary anticodons in the tRNA. A ribosome can accommodate two tRNAs at any one time while their amino acids are being linked together by a peptide bond (Figure 12.10).

As the bond between adjacent amino acids is formed, the ribosome simultaneously moves one triplet further along the messenger. The tRNA on the left (Figure 12.10) is then released, to be used over again. The next tRNA comes in at the right to pair with the newly positioned triplet and to add another amino acid to the growing polypeptide chain. As the ribosome moves 'down' the mRNA molecule, amino acids are joined to the growing chain at the rate of 15 per second. The complementary nature of the base-pairing between the mRNA codons and the anticodons of the tRNAs ensures that the transcribed message in the mRNA is faithfully translated into the correct sequence of amino acids in the polypeptide product.

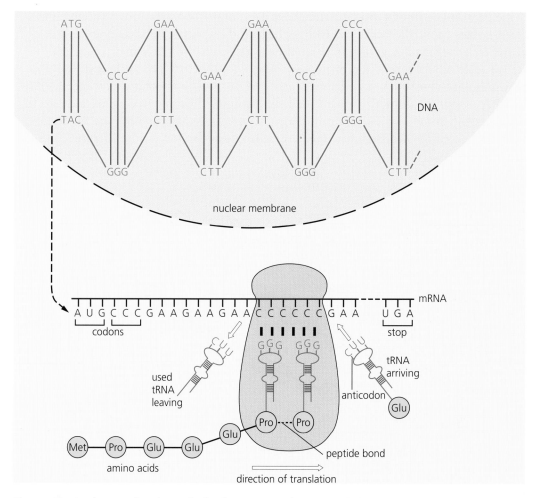

Figure 12.10 Diagram showing a single ribosome translating a coded message in mRNA into a polypeptide chain

Polypeptide chains usually begin with the amino acid methionine. This is coded for by the triplet **TAC** in the DNA. It is transcribed as the codon **AUG** in mRNA and recognised by tRNA with the anticodon **UAC**. The average length of a polypeptide is 333 amino acids. Their sequence is completed when the ribosome comes to one of three special 'stop signals' – **UAA**, **UGA** or **UAG**. The tRNAs that pair with the codons of these stop signals do not carry an amino acid.

The process by which the transcribed information carried in the base sequence of mRNA is used to produce a sequence of amino acids in a polypeptide chain is known as **translation**.

Once the polypeptide is synthesised, it dissociates from the ribosome and is released into the cytoplasm. There it may undergo some post-translational modification, such as folding or associating with other polypeptides, to form a functional protein.

Electron micrographs showing the translation of mRNA often reveal long chains of ribosomes (**polyribosomes**). It is thought that these occur because one mRNA strand is being simultaneously translated by a succession of ribosomes. Once the amino acids have been linked by peptide bonds, and their tRNAs released, the beginning of the mRNA strand is free, and can be used by another ribosome. As this second one moves along the strand the beginning will become free again and a third ribosome can come into play. A number of polypeptides, all of which are identical, can thus be synthesised from one mRNA molecule at the same time (Figure 12.11).

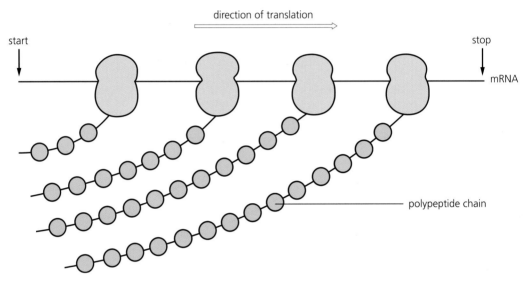

Figure 12.11 Diagram showing how several ribosomes can simultaneously translate the same mRNA strand

The flow of information

The question we posed earlier about how the genetic information in DNA may be used to determine the structure of proteins has now been answered. The answer may be summarised in terms of the *flow of information* (Figure 12.12).

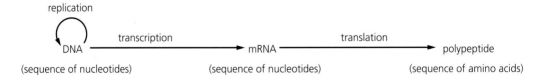

Figure 12.12

An important point to note here is that the flow of information takes place at all stages by the principle of complementary pairing of nucleotide bases.

Prokaryotes and eukaryotes

The flow of information takes place in the same general way in both prokaryotes and eukaryotes, but there are important differences between the two groups: these are described in Box 12.2.

Box 12.2 Flow of information in prokaryotes and eukaryotes

In the prokaryotes there is no nuclear membrane and therefore no need for the transcribed mRNA to be moved from the nucleus to the cytoplasm – it is already in the cytoplasm. In bacteria, and the other prokaryotes, translation and transcription are *coupled*. As the mRNAs are being transcribed from the DNA, the ribosomes attach to them and translation begins even before transcription is complete. In eukaryotes, as we have described, this is not the case: transcription is completed first and then the mRNA moves out of the nucleus, where it is translated. Furthermore, it is now known that in eukaryotes the mRNA transcripts are *modified* by the addition of 'caps' and 'tails', before they leave the nucleus (Figure 12.13). Before transcription is completed, a 'cap' consisting of a modified guanine is attached to the 5' end of the transcript by a bond that includes three phosphate groups. The guanine is modified by **methylation**; that is, the addition of extra methyl (CH_3) groups. The function of the cap is not fully understood, but it appears that it may act as a signal to promote translation.

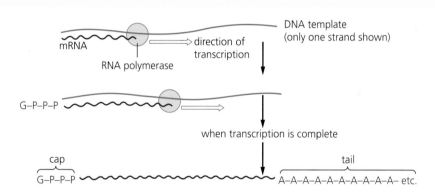

Figure 12.13 Diagram to illustrate the modification of mRNA transcripts in eukaryotes by the addition of 'caps' and 'tails'

Most mRNA transcripts in eukaryotes are also modified at the 3' end. Almost immediately after translation is completed, a 'tail' of a hundred or so adenine nucleotides is added. The adenines are collectively called poly-A. The poly-A tail is thought to act as a signal for export of the messenger out of the nucleus, and also to protect it from breakdown by enzymes in the cytoplasm. Messengers without a tail last only minutes in the cytoplasm, whilst those which have a tail may last for several days. (See also Box 12.3, page 151, for further details on processing of mRNA in eukaryotes.)

Deciphering the code

The genetic code has been completely deciphered and all 64 codons have been assigned to their respective amino acids or stop signals. This was done in the 1960s in a series of experiments devised by Nirenberg and Mathaei. They constructed synthetic mRNAs of simple known sequence, and then introduced these mRNAs into cell-free extracts of *E. coli* in a test tube. The cell-free extracts contained the necessary 'machinery' of protein synthesis; that is, ribosomes, tRNAs, enzymes and amino acids. Polypeptides were synthesised and isolated using this cell-free system. When a synthetic messenger comprising only uracil nucleotides (poly-U) was used, a polypeptide made entirely of phenylalanine residues was the result (Figure 12.14).

Figure 12.14

second base

		U	C	A	G	
first base	U	UUU ⎫ Phe UUC ⎭ UUA ⎫ Leu UUG ⎭	UCU ⎫ UCC ⎪ Ser UCA ⎬ UCG ⎭	UAU ⎫ Tyr UAC ⎭ UAA Stop UAG Stop	UGU ⎫ Cys UGC ⎭ UGA ⎱ Stop UGG ⎰ Trp	U C A G
	C	CUU ⎫ CUC ⎪ Leu CUA ⎬ CUG ⎭	CCU ⎫ CCC ⎪ Pro CCA ⎬ CCG ⎭	CAU ⎫ His CAC ⎭ CAA ⎫ Gln CAG ⎭	CGU ⎫ CGC ⎪ Arg CGA ⎬ CGG ⎭	U C A G
	A	AUU ⎫ AUC ⎬ Ile AUA ⎭ AUG Met	ACU ⎫ ACC ⎪ Thr ACA ⎬ ACG ⎭	AAU ⎫ Asn AAC ⎭ AAA ⎫ Lys AAG ⎭	AGU ⎫ Ser AGC ⎭ AGA ⎫ Arg AGG ⎭	U C A G
	G	GUU ⎫ GUC ⎪ Val GUA ⎬ GUG ⎭	GCU ⎫ GCC ⎪ Ala GCA ⎬ GCG ⎭	GAU ⎫ Asp GAC ⎭ GAA ⎫ Glu GAG ⎭	GGU ⎫ GGC ⎪ Gly GGA ⎬ GGG ⎭	U C A G

third base

Figure 12.15 The genetic code, given in terms of the mRNA codons (as is the convention)

The code for phenylalanine was therefore shown to be **UUU**. Similar tests, with poly-A and poly-C, revealed that **AAA** coded for lysine and **CCC** for proline. Poly-G didn't work: it folded up in a way that would not allow for translation. Synthetic messengers with mixtures of nucleotides in various ratios eventually revealed the assignments of all the codons, as given in Figure 12.15. In referring to the genetic code, we are therefore talking about the codons in mRNA: these are complementary to the sequences in the transcribed coding strand of DNA and identical with those in the non-transcribed strand.

Features of the code

It is not necessary to remember the assignments of the various codons, but there are certain features of the genetic code that should be noted.

1 Certain amino acids (methionine and tryptophan) are coded for by only one codon, but most are coded for by several (Figure 12.15). The code is therefore said to be **degenerate**, because it contains more information than is required; that is, there are more codons than there are amino acids. This degeneracy is mostly accounted for by variations in the third base of the codon: for example **GCU**, **GCC**, **GCA** and **GCG** all code for alanine. The family of tRNAs that carry alanine therefore recognise only the first two bases in the mRNA codon, and the third one is relatively unimportant.

2 Some triplets do not code for any amino acids: these are the stop codons, or chain-terminating codons, **UAA**, **UAG** and **UGA**. They play a vital role as signals to terminate translation and to bring an end to the formation of a polypeptide chain. The stop codons can be thought of as a form of punctuation that separates one gene from its neighbours.

3 The code is continuous and non-overlapping. If it were overlapping, so that the beginning of one triplet was also used for the end of the one preceding it, then a mutation (say a base-pair substitution) in one codon could affect more than just one amino acid in a polypeptide – and this is not the case.

4 The code is universal and the same codons are used to specify the same amino acids in all living organisms – with one exception. In the mitochondrial DNA of humans, and of yeast, certain chain-terminating codons are read as amino acid-specifying codons, and vice versa.

What is a gene?

Thus far we have only been able to envisage the gene as a particulate structure that we have referred to as a 'unit of heredity'. This definition was based on our observation of the way in which characters are inherited in crossing experiments. Patterns of inheritance could only be explained if the characters concerned were determined by particular unit structures, genes, which were present as allelic pairs at corresponding loci within homologous chromosomes.

The substance that makes up the chromosomes, and which carries information in genes, is DNA. Now that we know about the structure of DNA, and about the way it acts to determine characters, we can say what the gene is in precise biochemical terms.

A gene is a sequence of nucleotide pairs along a DNA molecule that codes for an RNA or polypeptide product. Both strands of the double helix make up the gene. Only one strand, the 'transcribed coding strand', contains the information that is directly utilised for RNA or amino acid assembly. The other strand, that is the 'non-transcribed' one, contains nucleotide sequences that are complementary to those of the coding strand, and is used for replication. The transcribed coding strand is not necessarily the *same* strand in all of the genes along a chromosome: it switches from one to the other. The structure of a gene is shown diagrammatically in Figure 12.16.

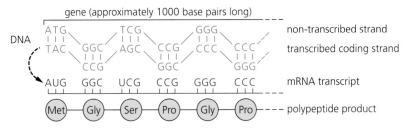

Figure 12.16 Diagram of a small part of a structural gene, showing the organisation of base triplets

Genes that code for polypeptides fall into two main classes. These are:
1 structural genes, which code for functional proteins (enzymes, hormones, components of cell structure, antibodies, storage proteins, and so on)
2 regulatory genes, which serve to control the activity of other genes.
Polypeptide chains vary in their length and so too do the genes that code for them. The average size of a polypeptide chain is 333 amino acids, so the average size of a structural gene is of the order of 1000 nucleotide pairs of DNA.

Transfer and ribosomal RNA molecules are also coded for by genes, but they are not synthesised in the same way that proteins are. The tRNA and rRNAs are made directly by transcription from the DNA, in the same way as mRNA. Transfer RNA molecules are about 80 nucleotides long, and the genes that code for them are therefore of corresponding length in terms of DNA. Ribosomal RNAs are of three sizes – 100 bases, 1500 bases and 3000 bases, and their genes consist of corresponding lengths of DNA nucleotides. The genes that code for rRNAs are present in multiple copies, and in eukaryotes they are localised at a special region in the chromosome called the secondary constriction, or nucleolus organiser.

Split genes and redundant DNA

It is impossible to give a complete account of the structure and action of DNA, and of genes, because new facts and surprises are emerging all the time as research into molecular genetics continues to race ahead. New techniques of genetic engineering (Chapters 18 and 19), first developed in the 1970s, have made it possible to study the structure and organisation of genes, and of chromosomes, in great detail. Individual genes, and large stretches of chromosomal DNA, can now be sequenced, and their nucleotide composition analysed and compared with their RNA transcripts and protein products (Chapter 18).

It turns out that most eukaryotic genes are 'split', and have far more DNA within them than is actually required to code for the amino acids in their protein products; that is, they contain stretches of **non-coding DNA**. It also transpires that large regions of the chromosomes do not appear to contain any genes at all: they are composed of stretches of repetitive DNA; that is, small sequences of bases that are present as millions of tandemly repeated copies. This repetitive (or redundant) DNA has no known function. Some of it is known to be transcribed, in certain organisms, but we have no idea what the protein products are (if any), or what purpose they serve within the cell. Further details are given in Box 12.3.

Box 12.3 Splicing of the primary transcript

In eukaryotes, as we have explained, transcription and translation are separate events. Transcription takes place within the nucleus and the mRNA then moves out into the cytoplasm for its translation.

It turns out that there is another event – a processing stage – that happens while the mRNA transcript is still within the nucleus. This is necessary because many eukaryote genes are 'split' and contain several stretches of additional DNA, called **introns**, over and above that required to specify the amino acids of their proteins. The coding sequences are known as **exons**. When transcription takes place all of the DNA bases are copied in the mRNA transcript. The processing stage then occurs and snips out the introns, by means of a splicing enzyme, to form the mature mRNA that moves to the cytoplasm for translation. Only the coding triplets in the exons are used to determine the amino acid sequence of the polypeptide (Figure 12.17, overleaf). The function of the DNA in the non-coding introns is not known. Splicing of the primary transcript takes place after the addition of the 'caps' and 'tails' mentioned earlier (see page 147).

Many gene sequences are composed mostly of introns. A gene associated with one form of muscular dystrophy in humans, for example, is about two million base-pairs long, but only 1200 of these actually code for protein (less than 1%). This coding DNA is distributed throughout the gene as 75 exons.

Introns form only one category of the redundant DNA that is a general feature of eukaryote genomes. Other kinds of non-coding DNA include spacer sequences that occur between certain genes, and short repetitive sequences of bases, too small to be genes, which are present as millions of repeated copies scattered throughout the chromosomes.

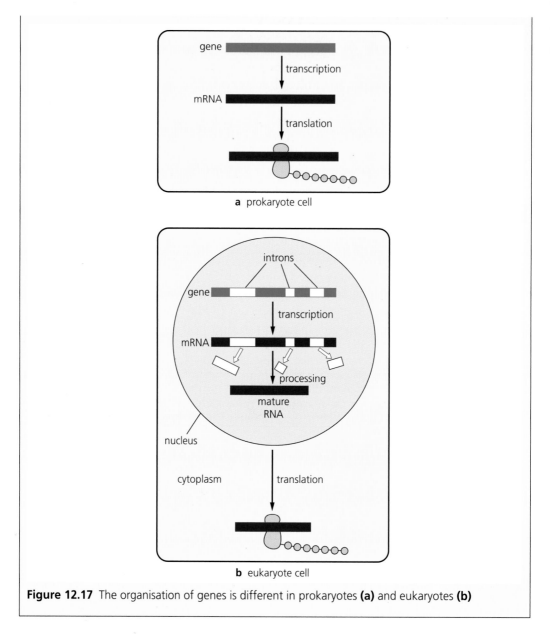

a prokaryote cell

b eukaryote cell

Figure 12.17 The organisation of genes is different in prokaryotes **(a)** and eukaryotes **(b)**

Genes and characters

In this chapter, we have now seen how a simple biochemical defect, such as an enzyme deficiency, can lead to a visible change in the phenotype of a character. This is evident in the metabolic disorders in humans and the auxotrophic mutants in *Neurospora*. Since the enzymes concerned are directly controlled by individual genes, the relation between the gene and the character is obvious. We say more on this matter in Chapter 13, after dealing with the subject of mutation.

The genome

We have seen that DNA contains far more than just genes. Much DNA is non-coding and never transcribed into messenger RNA. Other DNA within genes is transcribed but then removed from the mRNA before translation. The total DNA content of a cell is called its **genome**. The **nuclear genome** is the total DNA content of the haploid nucleus; that is, not including the genomes of cell organelles such as mitochondria and (in plants) chloroplasts. No one is certain why genomes contain so much apparently useless DNA. *Genomes* are species-specific – for example, the human genome – as compared to *genotypes*, which are specific to individuals.

Box 12.4 What makes a genome?

The size of the genome varies widely between species. Some differences make sense in terms of organism complexity; for example, humans have almost 250 times more DNA per haploid genome than yeast. Others, however, do not; for example, the fritillary plant (*Fritillaria assyriaca*) has over 40 times more DNA per haploid genome than humans, while the newt *Triturus cristatus* has over 5 times more. This apparent disparity between the amount of DNA per genome and the evolutionary complexity of species is called the '**C-value paradox**', and its occurrence caused geneticists to question the nature of the 'extra' DNA in species like the newt and fritillary.

By studying the genome compositions of different species, they discovered that there are three main classes of DNA sequence: 'unique and low copy sequence' DNA, 'middle-repetitive' DNA and 'highly-repetitive' DNA. They also realised that much DNA codes for nothing, and has little or no effect on the phenotype. The genomes of different species have the different types of DNA in different proportions.

'Unique and low copy sequence' DNA includes most gene-related sequences and may make up between 30% and 70% of the genome in most eukaryotes; for example, about 50% in humans and 70% in the fruit fly *Drosophila melanogaster*. In contrast, the genomes of viruses and prokaryotes are entirely composed of single-copy sequences.

'Middle-repetitive' DNA sequences are repeated from a few to 10^5 times per genome. There are many families of related sequences, including genes. Again, the proportion of this type of DNA varies between eukaryotes – it makes up about 12% of the *Drosophila* genome and at least 40% of the human genome. The genes for tRNA and rRNA come into this category, as do those for histone proteins. Each histone is encoded by a separate gene, repeated about 10 times per genome in some birds, 20 times in mammals, 100 times in *Drosophila*, and up to 600 in some species of sea urchin.

'Highly-repetitive' DNA mostly consists of short sequences – up to 300 base-pairs, although most are much shorter – repeated in excess of 10^5 times per genome, including the tandemly repeated DNA already referred to. It may constitute up to about 45% of the genome, depending on the species. Highly-repetitive DNA can often be seen as deeply staining regions near the centromeres and telomeres (ends) in metaphase chromosome preparations – in 'heterochromatic blocks'. Few, if any, genes are located in this **heterochromatin**, which may remain highly condensed throughout the cell cycle, and is genetically inert compared with the gene-containing **euchromatin**.

Summary

◆ Studies on metabolic disorders in humans, and the biochemical genetics of *Neurospora*, led to the idea that genes act by determining the structure of enzymes, and thereby control all the biochemical activities of the cell.

◆ It was subsequently shown that genes determine the structure of all proteins, including enzymes, and that one gene is responsible for specifying the sequence of amino acids of one polypeptide chain.

◆ The information in the DNA is encoded in such a way that one triplet of bases carries the information specifying one amino acid. When the information is used, it is first of all transcribed into a molecule of single-stranded messenger RNA, and then translated into protein through the involvement of ribosomal and transfer RNA. The process is known as protein synthesis.

◆ The ribosomal and transfer RNAs are themselves transcribed directly from DNA. The gene is thus defined as a sequence of nucleotides of DNA that codes for an RNA or protein product.

◆ In eukaryotes there is a large amount of non-coding, redundant DNA, the function of which is unknown.

Further reading

1 G. W. Beadle (1948) 'The genes of men and moulds', *Scientific American* **179**, No. 9, 30–40.

2 P. Chambon (1981) 'Split genes', *Scientific American* **244,** No. 5, 60–71.

3 F. H. C. Crick (1962) 'The genetic code', *Scientific American* **207,** No. 4, 66–74.

4 F. H. C. Crick (1966) 'The genetic code: III', *Scientific American* **215**, No. 4, 55–62.

5 M. W. Nirenberg (1963) 'The genetic code: II', *Scientific American* **208**, No. 3, 80–94.

6 W. F. Doolittle and C. Sapienza (1980) 'Selfish genes, the phenotype paradigm and evolution', *Nature* **284**, 601–3.

7 L. E. Orgel and F. H. C. Crick (1980) 'Selfish DNA: the ultimate parasite', *Nature* **284**, 604–7.

8 W. S. Clug and M. R. Cummings (2000) *Concepts of Genetics*. Sixth edition. Prentice Hall, Inc. Website: **http://cw.prenhall.com/bookbind/pubbooks/klug3/**

9 *DNA from the beginning*. Animated primer on the basics of DNA, genes and heredity: **http://vector.cshl.org/dnaftb**

10 *The Natural History of Genes*. Classroom and home activities, including how to extract DNA: **http://glsc.genetics.utah.edu**

Questions

1 Give an illustrated account of the structure of
a DNA
b chromosomes.

2 a Compare the structures of RNA and DNA.
b Describe the part nucleic acids play in the synthesis of proteins.

3 a Explain how the genetic message is encoded in DNA.
b How is the message **(i)** transcribed, and **(ii)** translated?

4 Describe how DNA is replicated during interphase.

5 a Which of the four major bases found in DNA are pyrimidines and which are purines?
b Which bases in DNA pair with which?
c Which four major bases are found in RNA?
d What are the sugars found in RNA and DNA? State which is found in which.
e What is a nucleotide?
f What types of RNA are involved in protein synthesis?

6 About 28% of the nucleotides of herring DNA contain adenine. What percentage of nucleotides would you expect to contain guanine?

7 The diagram below shows part of a molecule of messenger RNA.

U A C G G A C G A U A A C C U G G A

a What is a codon?
b How many codons are shown in the diagram?
c tRNA molecules carry a complementary base sequence for a particular codon. Write down the complementary sequences for each of the codons shown in the diagram.
d State *two* ways in which the structure of a tRNA molecule differs from that of a DNA molecule.
e Describe the role of tRNA in protein synthesis.

8 The diagram below represents the structure of the basic chemical unit from which RNA and DNA are formed.

a Name the chemical unit in the diagram.
b State the names of the components labelled A and B.
c Which of the bases found in RNA is not usually found in DNA? What base usually replaces that base in DNA?
d Which base pairs with cytosine?

9 Give an account of the structure and replication of deoxyribonucleic acid (DNA).   [10]

London AS/A Biology and Human Biology Module Test B/HB1 June 1999 Q8

10 a Name the type of bond that holds together the two strands of nucleotides in a DNA molecule.   [1]

Genetic drugs are short sequences of nucleotides. They act by binding to selected sites on DNA or mRNA molecules and preventing the synthesis of disease-related proteins. There are two types:

- *Triplex drugs* are made from DNA nucleotides and bind to the DNA forming a three-stranded helix.
- *Antisense drugs* are made from RNA nucleotides and bind to mRNA.

b Name the process in protein synthesis that will be inhibited by:

(i) triplex drugs   [1]

(ii) antisense drugs.   [1]

c The table shows the sequence of bases on part of a molecule of mRNA.

Base sequence on coding strand of DNA									
Base sequence on mRNA	A	C	G	U	U	A	G	C	U
Base sequence on antisense drug									

Copy and complete the table to show:

(i) the base sequence on the corresponding part of the coding strand of a molecule of DNA   [1]

(ii) the base sequence on the antisense drug that binds to this mRNA.   [1]

AEB AS/A Biology Module Paper 2 June 1998 Q1

11 The table shows the relative amounts of four bases in DNA taken from cells of three organisms.

	Nitrogenous base (relative amounts)			
Cellular source of DNA	**Adenine**	**Guanine**	**Cytosine**	**Thymine**
rat bone marrow	28.6	21.4	21.5	28.4
wheat grain	27.3	22.7	22.9	27.1
yeast	31.3	18.7	17.1	32.9

a (i) Explain why, in a rat bone marrow cell, the amount of adenine is approximately equal to the amount of thymine.   [2]

(ii) If DNA was taken from other tissues of this rat, explain why it would have the same base composition as that taken from the bone marrow cells.   [2]

(iii) Explain why the relative amounts of each base are different in all three organisms.   [2]

In the process of protein synthesis, information in DNA is used to construct polypeptides.

b (i) Explain the difference between *transcription* and *translation*.   [2]

(ii) Describe the role of transfer RNA in protein synthesis.   [4]

OCR (Cambridge Modular) AS/A Sciences: Central Concepts in Biology November 1999
Section A Q1

13

Gene mutation

We have now seen how the structure of DNA allows for the self-replication, and for the encoding and utilisation, of genetic information during gene action. The processes of replication, transcription and translation occur with an extraordinarily high level of fidelity. This is due to the very specific way in which the nucleotide bases bond together as complementary pairs. The stability of DNA, the accuracy with which it is replicated, and the way the information encoded within it is used, are the essence of heredity. This constancy enables species to maintain their genetic programs and to preserve their identity over countless cycles of development and reproduction.

But DNA replication is not perfect. If it were, diploid organisms would have two identical copies of each gene at all their different loci, and every individual of a species would be of the same genotype. There would be no genetic variation and no capacity to adapt to long term changes in the environment. By the same token there would be no character differences with which the geneticist could work and no alleles with which to study the inheritance, structure and action of the genetic material.

The capacity for change, and for heritable variation, comes about by mutation.

A **mutation** is a sudden heritable change in the genetic material. Gene mutations (or point mutations) result from changes occurring within single genes, and they give rise to new alleles. In referring to 'change', we mean some departure from the normal form of a gene, or from the form normally present in an experimental laboratory strain of an organism that we use as our standard type.

How do gene mutations arise?

Gene mutations can arise spontaneously or they can be induced.

Spontaneous mutations result from errors in the replication of DNA. We have only mentioned one of the enzymes involved in this process, DNA polymerase, but in reality replication requires a vast number of different enzymes. These enzymes make very rare mistakes and this leads to changes in the base composition of DNA.

Induced mutations are changes in DNA that are caused by the effects of 'mutagens'. Changes in the base composition of DNA, spontaneous or induced, constitute the molecular basis of gene mutation.

Molecular basis of gene mutation

There are three main ways in which the base composition of the DNA of a gene may be altered by mutation – these are shown in Figure 13.1. The changes may involve just one, or more than one, nucleotide pair. These changes can be detected because they affect the information carried in the genetic code. If we recall the pathway of information flow given in Chapter 12 (see page 146), it is obvious that a change in the nucleotide sequence of a gene can result in a corresponding change in the amino acid sequence of its polypeptide product. Looking at the changes in terms of gene action and protein products, we can classify mutations in several categories.

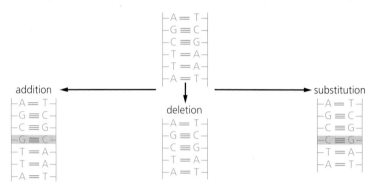

Figure 13.1 Alterations in the base composition of DNA causing gene mutation

Same-sense mutations

A **same-sense mutation** is a base substitution in a DNA triplet that does not alter the amino acid sequence of a polypeptide.

These mutations can happen because the genetic code is degenerate. The majority of amino acids are coded for by several mRNA codons, which vary in their third base (see page 148). Proline, for instance, is coded for by **CCU, CCC, CCA** and **CCG**. If a mutation in the third base of a DNA triplet changes an mRNA codon from **CCA** to **CCC,** it will not make any difference to the peptide that is formed.

Mis-sense mutations

A **mis-sense mutation** is a base-pair substitution in a gene that results in one amino acid being changed for another one at a particular place in a polypeptide.

For example, if a mutation in DNA alters a mRNA codon from **UUU** to **UGU** then, at a corresponding place in the polypeptide product, phenylalanine will be replaced by cysteine. Such changes may have relatively little effect upon the function of the protein concerned, or they may be quite serious as in the case of sickle-cell anaemia.

Nonsense mutations

A **nonsense mutation** is any mutation (substitution, addition or deletion) that changes an amino acid-specifying codon into a chain-terminating codon (**UAG, UAA** or **UGA** in mRNA).

This results in the premature termination of a polypeptide chain. The way in which a substitution can cause a nonsense mutation is illustrated in Figure 13.2.

Normal code in mRNA (direction of reading⇒)

AUG UAU CUG AUU UCA AAA CCC . . .   etc. (say 300 codons)
Met   Tyr   Leu   Ile   Ser   Lys   Pro   —   etc. (300 amino acids)

Substitution

AUG UAU CUG AUU UGA AAA CCC . . .   etc.
Met   Tyr   Leu   Ile   STOP

Figure 13.2 A nonsense mutation results in the premature termination of the polypeptide product

Frameshift mutations

A **frameshift mutation** is a mutation resulting from the addition or deletion of one or more nucleotides, other than in multiples of three, that causes the gene to be 'misread'.

The **reading frame** describes the way in which the code in mRNA is read as triplets by the ribosome. The insertion or deletion of bases, other than in multiples of three, results in a shift in the reading frame and creates an entirely new sequence of codons – some of which are nonsense codons. The gene product is therefore non-functional because it has the wrong sequence of amino acids in the polypeptide and the chains are also terminated prematurely (Figure 13.3a). The insertion or deletion of three bases does not affect the reading frame, but it does result in the inclusion or loss of one amino acid from the polypeptide (Figure 13.3b). It was Francis Crick who first discovered the effect of adding or deleting bases in threes and this gave him the vital evidence that the code was organised as *triplets* of bases.

In the above account, we have concentrated on genes that code for polypeptides. It should be mentioned that mutations occur in genes coding for rRNAs and tRNAs as well.

Normal code in mRNA (direction of reading⇒)

AUG AAU CUC UUU GUG GGU AGG . . .   etc.
Met   Asn   Leu   Phe   Val   Gly   Arg   —   etc.

a insertion of one base – frameshift mutation

AUG AAU GCU CUU UGU GGG UAG G . . .   etc.
Met   Asn   Ala   Leu   Cys   Gly   STOP   —   etc.

shift in reading frame

b insertion of three bases – mis-sense mutation

AUG AAU GUA CUC UUU GUG GGU AGG . . .   etc.
Met   Asn   Asp   Leu   Phe   Val   Gly   Arg   —   etc.

Figure 13.3 Frameshift mutations result from the insertion or deletion of a number of nucleotides, other than multiples of three **(a)**. In most instances the polypeptide product is completely non-functional. When three bases are added, or deleted, only one amino acid is affected and the reading frame remains in phase **(b)**

Somatic cells and gametes

It is important before proceeding any further to distinguish between mutations that arise in somatic cells and those that occur in the germ line or in the gametes themselves – because the consequences are quite different.

Somatic cell mutations only affect a few cells in an individual higher organism, and the majority of the tissues are unaffected and are of normal phenotype. Such mutations may be quite trivial in their effects, and in a sexually reproducing species they are *not* inherited. Some of them can be serious to the individuals concerned if they result, for instance, in a detrimental condition such as a malignant tumour in an animal.

Mutations in the germ line, or in the gametes, are of much greater significance since they will be present in *all* of the cells of the progeny that arise from the affected gametes. They will be transmitted to future generations and may make a contribution to the genetic variability of the species (Chapter 16).

Somatic cell mutations can be of importance in asexual reproduction if the cells or tissues involved happen to be in that part of the organism that becomes detached to give a new individual; for example, asexual spores in fungi.

Gene mutations and character differences

Character differences result from gene mutations. In this section, we will first of all consider how a mutation affects a character. We will then explain why it is that most mutations are deleterious and recessive. Finally we will itemise the kinds of effects that they have upon the phenotype.

How do mutations cause character differences?

In the present context, we can consider that the expression of genetic information involves two steps. The first one is the transcription and translation of the information into a polypeptide. The second is concerned with the action of the polypeptide (Figure 13.4).

Figure 13.4

When a mutation takes place, the information in the gene is changed, and it is then expressed differently (Figure 13.5).

Figure 13.5

In a diploid organism, there are two alleles being expressed in the same cell. When they are identical (**AA** or **aa**), we have no difficulty in understanding how they contribute to the character. When they are different (**Aa**), we have to explain how they act, or interact, together. The details of the explanation are given in Box 13.1. We also have to remember at this stage that the actual phenotype we observe depends upon the way in which the genetic effects interact with the environment.

We have already discussed the mutation that causes sickle-cell anaemia (Chapter 12, page 140). In this case, we know exactly what biochemical changes are involved in the mutant allele (single base-pair substitution), how it alters the β polypeptide product (single amino acid substitution) and what the consequences are to the phenotype. This is one of the few mutations that we understand completely.

Box 13.1 Mutations and character differences

The simplest situation with which to explain how mutations affect characters is that in which the change results in complete loss of enzyme activity. This could arise either because the polypeptide is not formed, or because it is altered in such a way that it cannot function normally. The change can be symbolised as **A → a**. **A** is the normal form of the gene coding for an enzyme that converts a precursor substance (call it X) into, for example, the purple-coloured pigment of a flower; **a** is the recessive allele that is unable to give rise to a functional enzyme. In the dominant homozygote (**AA**), both alleles will produce the enzyme, giving a normal purple-flowered phenotype. The recessive homozygote (**aa**) will not be able to metabolise the precursor and the flower will exhibit a white mutant phenotype. The heterozygote (**Aa**) shows full dominance because the normal allele produces sufficient enzyme to make all of the pigment required for purple flowers (Figure 13.6).

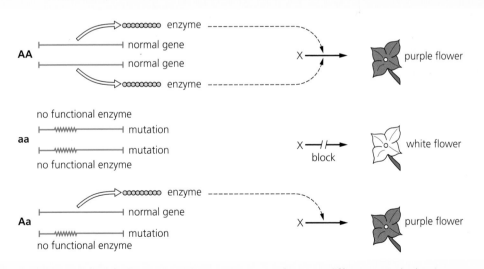

Figure 13.6 Explanation of how a mutation can cause a character difference, and what is meant by 'dominant' and 'recessive' in molecular terms. **A** is the normal dominant allele and **a** is the recessive mutant allele

Another situation is that involving incomplete dominance. Suppose we are again dealing with a gene controlling a flower colour phenotype, such that **RR** gives red flowers, **Rr** gives pink and **rr** is white. To explain this effect, we simply have to postulate that the normal allele (**R**) in the heterozygote produces insufficient enzyme to synthesise the full complement of red pigment.

The co-dominance that we find in the ABO blood group system in humans (Chapter 9) is a kind of protein polymorphism, which is not evident in the visible phenotype. This character is explained because the heterozygote (I^AI^B) has two variants of a normal gene both of which produce a product (Figure 13.7).

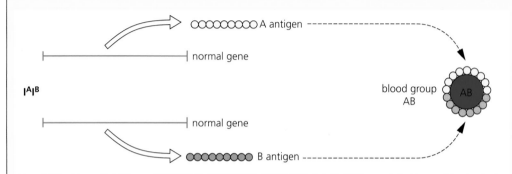

Figure 13.7 The alleles I^A and I^B are co-dominant – both produce a polypeptide product

We have confined ourselves here to a few simple examples. There are many other complications that could be considered, including those where we have interactions between two genes controlling the same character. We must also remember that mutations do not necessarily have all-or-nothing effects. Their consequences to the phenotype depend upon the kind of change that has taken place within the gene and the way in which the amino acids are altered in the gene product.

Mis-sense mutations are generally less serious than nonsense and frameshift mutations for the reasons given earlier in this chapter.

Why are mutations mainly recessive and deleterious?

Each gene that codes for a polypeptide is made up of a certain sequence of nucleotide pairs of DNA. The code is organised in triplets of bases that are transcribed into a corresponding sequence of amino acids in the protein product. The correct sequence of amino acids is essential for the protein to function properly (although *some* minor changes can be tolerated).

Since a mutation is a random change in the structure of a gene, it follows that it will result in a corresponding random change in the structure of the protein that it determines. As a result, the protein will suffer some impairment to its function (mis-sense mutation), or else it will not be produced at all (nonsense mutation). By analogy, if we were to cause a random change in the circuitry of a modern television receiver by poking about in the back with a screwdriver, we would most likely damage the quality of the picture, or else lose it completely. It is most unlikely that we would bring about any improvement! Mutations are therefore mainly deleterious because they upset the normal working of a gene. They are usually recessive for the simple reason that, in a heterozygote, a normal allele is present as well. The mutant one is hidden by virtue of the fact that it does nothing, or determines a non-functional protein (Box 13.1).

Forward and back mutations

The change from a normal to a mutant form of a gene is known as a 'forward mutation' ($A \rightarrow a$). A change in the opposite direction, from the mutant back to the normal form, is a 'back (or reverse) mutation' ($a \rightarrow A$). The forward mutation rate of a gene is much higher than the back mutation rate. The reason for this is that a forward mutation (say a substitution) involves a random change anywhere within the 1000 base-pairs of the gene. To reverse this mistake, the second random change must involve the same single base-pair (or another that compensates for the change), rather than *any* of the pairs, and the chances of this happening are clearly much lower.

Types of mutations

Mutations are classified into a number of different types. This is purely for convenience so that we can discuss them and write about them in a way that relates to the kind of effects they have upon the phenotype.

1 **Morphological mutations**: These are the most obvious types and they affect the form (= 'morph') of an organism. Included in this category are all of Mendel's pea characters and many of those that we have encountered in *Drosophila* (white eyes, vestigial wings, curled wings, ebony body, and so on). We also include here characters to do with appearance in humans (eye colour, albinism), comb form in fowl and any other characters that affect the size, shape or colour of an organism.

2 **Lethal mutations**: Lethal mutations kill an organism. An albino plant, which is lacking in chlorophyll, is a good example. In many cases though, we have no idea why the mutation causes death – we simply classify it by its effect.

3 **Biochemical mutations**: We have dealt with some of these in fungi and humans (see pages 137 and 138). They are identified by the loss of, or defect in, some specific biochemical function of a cell.

4 **Conditional mutations**: These are mutations that are expressed only under certain conditions; for example, the development of coat colour pattern in the Himalayan rabbit depends on temperature.

5 **Regulatory mutations**: Mutations in genes that control the activities of other genes.

Detection of mutations

Mutations in haploid organisms are easily detected, because there are no complications due to dominance and they are expressed immediately in the phenotype. In diploids, a dominant mutation (**A**) that arises in a gamete will also show up straight away in the progeny in which it is inherited. A recessive mutation (**b**), on the other hand, will not be apparent in the first generation (unless it is sex-linked) (Figure 13.8).

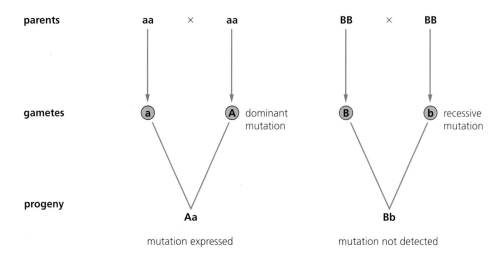

Figure 13.8

The recessive mutation may go undetected for several generations, even if it is lethal, until two heterozygotes mate together. It will then segregate out as a homozygote (**bb**).

In higher organisms, it is often difficult to find mutations unless they are very distinctive in appearance and easily spotted by eye – for the simple reason that they are rare (see below). Microbes, on the other hand, are much easier to deal with. To detect mutations in bacteria that give rise to resistance to antibiotics, for instance, it is simply necessary to grow up several million cells on agar medium containing the antibiotic (for example, penicillin). All the normal cells are killed and any colonies that survive are derived from mutant individuals with penicillin resistance. Where a *selective* system of this kind is used, the odd mutant can be screened out from amongst millions of normal cells.

At one time, it was suggested that such screening procedures were not actually detecting spontaneous mutations, but that the antibiotic substances themselves were causing the cells to mutate. This idea was disproved by Joshua and Esther Lederberg who devised a simple experiment to show that mutations giving resistance to certain antibiotics were present in the bacterial cultures, whether they were growing on the antibiotics or not. In other words, they confirmed that mutations are *random changes* in the genetic material: the details are given in Box 13.2.

Box 13.2 A replica-plating test to show that mutation is a random process

In 1952, Joshua and Esther Lederberg devised a simple test to show that mutations arise in bacteria by random processes and are not induced in response to a selective agent such as an antibiotic. They prepared a master plate containing several million colonies of *E. coli* on a complete agar medium. This medium contained no antibiotics. A velvet pad was then used to press onto the surface of the culture and to transfer a replica of the colony formation onto fresh plates of selective agar containing the antibiotic. A few colonies with antibiotic resistance grew on the new plates. Because of the way in which the colonies had been replica-plated it was possible to see that the *same* colonies were growing on each of the freshly inoculated plates. In other words, the mutations must have arisen spontaneously as random changes on the original master plate, and not in response to the antibiotics – otherwise they would not have been in identical places on the replicas. The experiment is illustrated in Figure 13.9.

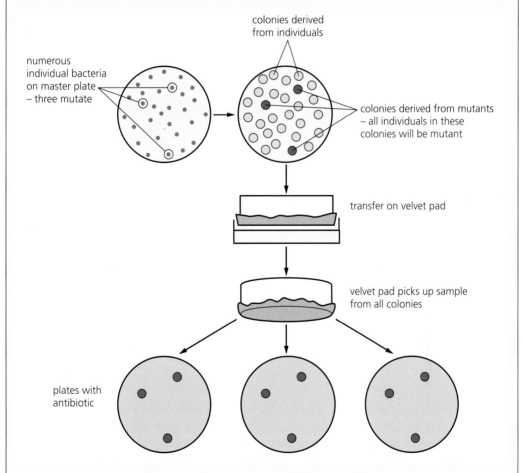

Figure 13.9 Some cells from each colony will get transferred to each of the three replica plates. Only those resistant to the antibiotic will grow

Mutation rates

Mutation rate is defined as the number of mutations in a given gene per unit of time, where time is measured as cell divisions. Mutation rates are difficult to measure in most organisms (except for viruses, bacteria and cell cultures) because we cannot determine the number of divisions that have given rise to a population of cells. In practice, therefore, and especially in higher organisms, we use *mutation frequency* as an estimate of mutation rate.

Mutation frequency is simply the frequency with which a mutation is found in a sample of cells or individuals. It is normally expressed as the number of mutations per million gametes, and if we make the appropriate testcrosses this is not too difficult to estimate (Table 13.1).

Table 13.1 Estimates of spontaneous mutation frequencies for genes in various organisms

Organism	Gene or character	Mutations per 1 000 000 gametes
Diplococcus pneumoniae	penicillin resistance	0.1†
E. coli	resistance to bacteriophage T1	0.03†
Drosophila	ebony body	20
	eyeless	60
	white eye	40
mouse	albino	10
maize	colourless aleurone (outer seed layer)	492
	purple seeds	11
	sugary seeds	2.4
	shrunken seeds	1.2
humans	muscular dystrophy	4–10*
	haemophilia	2–4*

* Range of values based on several estimates from pedigree records
† Based on cell counts

As far as we know, all genes mutate. The rate at which it happens spontaneously, under natural conditions, seems to vary quite widely between different organisms and different genes. In higher organisms, the frequency ranges between one in a million gametes (that is, 1×10^{-6}) to 6×10^{-5}, leaving aside the exceptionally high value of about 5×10^{-4} for the **R** locus in maize. In bacteria, the frequencies are lower (1×10^{-7} to 3×10^{-8}) (Table 13.1).

A number of genetic and environmental factors affect mutation rates.

Factors affecting mutation rates

Different genes mutate at different rates

A given gene has a fixed mutation rate under standard conditions, but these rates vary from one gene to another. One reason for this could be the difference in sizes of different genes – the larger the gene the greater the chance of a mutation.

Background genotype

A given gene may have different rates of mutation in different genetic backgrounds. The colour gene mutation **R** → **r** in maize happens three times more frequently in the Cornel strain than it does in the Columbia strain.

Transposable elements

Some genetic elements can cause other genes to mutate at an enhanced rate. One such case is that of the controlling elements discovered in maize by Barbara McClintock. These are short sequences of DNA bases that can move about (transpose) from one part of the chromosome to another! They are usually called **transposable elements** (or **transposons**) and when they insert themselves into a gene they upset its structure and cause a mutation.

Transposable elements are also well known in *Drosophila* and *E. coli*. Their existence was not believed for a long time because they don't fit very well with the chromosome theory of heredity.

Temperature

Spontaneous mutation rates can sometimes be altered by keeping organisms at higher or lower temperatures than normal.

Mutagens

A number of environmental agents, known as **mutagens** (radiation and chemicals), cause genes to mutate at rates that are much higher than their spontaneous level. Mutations that arise in this way are known as induced mutations.

Induced mutations are important for several reasons: they provide a useful tool for the geneticist; they may be useful to humans in relation to plant and animal breeding (Chapter 17); and the mutagens that cause them constitute an environmental hazard because they damage the genetic material of all living organisms.

In 1927, H. J. Müller found out that X-rays increase the mutation rate in *Drosophila*. He was later awarded the Nobel prize for this important discovery. Independently, and at about the same time, Lewis Stadler demonstrated that X-rays induce mutations in maize. Since then, a large number of different mutagens have become known. We can place them in three categories, as outlined below.

1 Ionising radiations

This category includes X-rays, cosmic rays and radiation coming from various radio-active sources; for example, alpha (α), beta (β) and gamma (γ) rays. They damage the genetic material directly by breaking it up, or else they break up other molecules (for example, water), which then become reactive and can damage the DNA indirectly.

With this kind of radiation there is a *linear* relationship between the induced mutation rate and the dose of radiation given (Figure 13.10). The graph can therefore be extrapolated back through the origin and we can conclude that radiation damage will be occurring even at levels that are too low to be detected experimentally. In other words, there is no such thing as a 'safe' dose. What is more, the effects are *cumulative*, at least to some extent, so that a high dose can be given in the form of several small doses over a period of time.

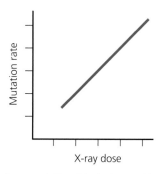

Figure 13.10 Graph showing the linear relationship between X-ray dose and induced mutation rate

2 Non-ionising radiations, that is ultraviolet light (UV)

UV light is less penetrating than ionising radiations and acts in a different way. It is absorbed by the purine and pyrimidine bases in DNA and modifies them in various ways. One of the main effects is to produce **thymine dimers**. This occurs when two adjacent thymine molecules in one of the DNA strands become linked together (**T^T**). The dimer forms a bulge in the helix and this disrupts the bonds between the thymine molecules in the dimer and their complementary adenines (**AA**) in the opposite strand. In the presence of visible light, many of these dimers are removed by repair enzymes that make good the damage.

There are two interesting aspects of UV-induced damage that we should mention in relation to the **light repair system**. The first is that when UV light is used to kill bacteria, the cells that have been treated should be kept in the dark. The second aspect concerns a skin cancer in humans known as xeroderma pigmentosum. This cancer is induced by the UV rays in sunlight and is found in individuals who have a mutation in a gene that determines one of the repair enzymes. The majority of us can tolerate sunbathing because of the highly effective enzyme repair system that makes good the damage we inflict upon ourselves.

3 Chemical mutagens

Chemical **mutagenesis** was discovered by Charlotte Auerbach and J. M. Robson in their experiments with mustard gas during World War II. Their work was classified as secret for several years. Mustard gas is an alkylating agent that affects the base guanine. Alkylating agents can change the pairing specificity of guanine, and so induce a base-pair substitution; they can cause it to become unstable so that it is released from the DNA leaving a gap, which may then be filled by another base.

There are numerous other chemical mutagens. Many of them are 'base analogues': they have structures very similar to the normal bases and become incorporated into the DNA in place of them during replication. They act through a process known as 'tautomerism', which can cause a change in their pairing specificity. An example is 5-bromouracil (5-BU), which is an analogue of thymine. It has a bromine atom at the ^5C position in place of the methyl group (CH_3) in thymine. This bromine atom causes the molecule to change spontaneously from one state to another by redistribution of its electrons: the change in state is known as a **tautomeric shift**. In the *keto* state, BU behaves like thymine and is complementary to adenine during replication (forming two bonds); whereas in its rare *enol* state it forms three bonds and pairs with guanine. This change in pairing specificity can lead to base-pairing substitution during DNA replication, and therefore to a mutation (Figure 13.11).

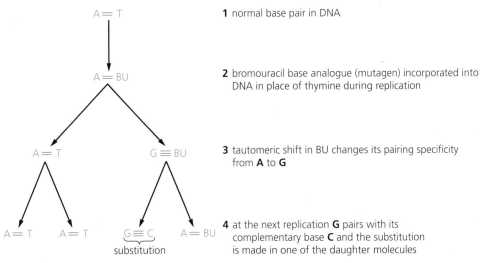

Figure 13.11

Fate of mutations

The question of what happens to gene mutations once they have arisen in a natural population is dealt with in Chapter 20. Generally speaking, most mutations are unfavourable for the reasons given (see page 162), and they are quickly eliminated by natural selection. Recessives in diploids may remain 'hidden' in heterozygotes for several generations before they segregate out as homozygotes. Each population therefore carries a genetic load of deleterious recessive alleles. Mutations that improve fitness will be favoured by selection and will increase in frequency, so that one allele may gradually replace another.

Evolution depends upon the tiny minority of mutations that are not harmful – although what is harmful in one environment may well turn out to be beneficial in another one at a different time. The importance of mutations in the genetic system of variation is discussed further in Chapter 15.

Summary

◆ DNA is a highly stable molecule, but rare errors in its replication produce heritable changes known as gene mutations.

◆ These random changes arise spontaneously by the substitution, addition or deletion of nucleotide bases in DNA.

◆ Mutations that occur in the gametes are more important than those that arise in somatic cells because they are transmitted to future generations.

◆ Gene mutations cause a variety of effects upon the phenotype: most are deleterious and recessive. The way in which they affect characters can be simply explained in terms of the biochemistry of gene action.

◆ Mutations are much easier to detect in micro-organisms than in higher plants and animals – this is because they are haploid and several millions can be screened using selective procedures to pick out only the mutants.

◆ Genes mutate spontaneously at very low rates. The rate can be increased by the use of physical and chemical mutagens.

◆ Mutations are important in experimental genetics and as a source of variation in natural populations.

Further reading

1 D. N. Cooper and M. Krawczak (1993) *Human Gene Mutation*. BIOS Scientific Publishers.
2 E. C. Friedberg, G. C. Walker and W. Siede (1995) *DNA repair and mutagenesis*. ASM Press.

Questions

1 Write an essay on mutations and mutagenesis.

2 **a** What is meant by *mutation*?
  b Name three agents that induce mutation.
  c Why are micro-organisms like bacteria frequently used to study mutations?
  d Why is mutation important in evolution?

3 **a** What is a codon?
  b Explain what is meant by the phrase 'the genetic code is degenerate'. Why does this sometimes lead to mutations having no effect on phenotype?
  c Why are most mutations recessive and deleterious?
  d What is a point mutation?

4 Flower colour in sweet peas is controlled by three genes, each with two pairs of alleles, **A** and **a**, **B** and **b**, and **C** and **c**. The dominant genes each produce an enzyme involved in pigment production. Precursor molecules are converted into an intermediate substance by enzyme **A**. Both the precursor and intermediate molecules are white. The enzyme produced by **B** converts intermediate molecules into blue pigment. The enzyme produced by **C** converts intermediate molecules into red pigment. The possible flower colours are red, blue, white and various shades of purple.
  a Which genotypes would you expect to have white flowers?
  b Which genotypes will have red flowers?
  c Which genotypes will have blue flowers?
  d Which genotypes would you expect to have purple flowers? Explain why the shade of purple might vary.
  e Explain why the mutations are recessive.
  f A plant of genotype **AABbCc** is self-pollinated. Explain, with the aid of a diagram, what the genotypes and phenotypes of the offspring will be. In what proportions are the possible phenotypes expected to occur?

5 The diagram shows the pathway by which phenylalanine is normally metabolised.

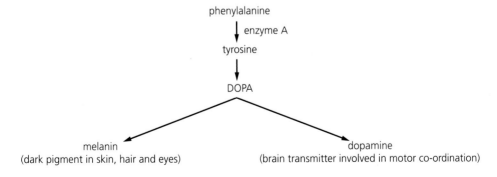

Phenylketonuria (PKU) is a condition which results from the absence of enzyme A. People with PKU are homozygous for a recessive allele which fails to produce this enzyme.

a Use the information shown in the diagram to give *one* symptom you might expect to be visible in a person who inherits PKU. [1]

b Explain how gene mutation may result in an allele which fails to produce a functional enzyme. [3]

c (i) A child with PKU was born to two unaffected parents. Copy and complete the genetic diagram to show how this is possible. [2]

parental phenotypes   _____   ×   _____

parental genotypes   _____   _____

genotypes of gametes   _____   _____

genotypes of children   _____

phenotypes of children   _____

(ii) What is the probability that a second child born to these parents will have PKU? [1]

AQA (NEAB) AS/A Biology: Continuity of Life (BY02) Module Test June 1999 Q4

6 a Explain why the genetic code for an amino acid is a three-base code rather than a two-base code. [3]

The table shows some of the DNA triplet codes together with their corresponding amino acids.

DNA triplet	Amino acid	Standard abbreviation for amino acid
AAA	phenylalanine	Phe
AAC	leucine	Leu
AAG	phenylalanine	Phe
AAT	leucine	Leu
CAT	valine	Val
CCA	glycine	Gly
CCG	glycine	Gly
CTT	glutamic acid	Glu
GCC	arginine	Arg
GTA	histidine	His
TCG	serine	Ser
TCT	arginine	Arg
TGC	threonine	Thr
TGT	threonine	Thr
TTT	lysine	Lys
ACT	nonsense/stop	

A section of the DNA template strand is represented below.

– C A T C C A A A T T G T T G C C C G –

b (i) Write the mRNA codon for the first DNA triplet of this section of the strand. [1]
(ii) Using the standard abbreviations given in the table write out the sequence of
amino acids coded by the section of DNA. [2]
It is possible for the DNA template strand to become changed or mutated in various
ways. Some of these mutations are shown below.

original:
– C A T C C A A A T T G T T G C C C G –

mutation 1:
– C A T C C A A A T T C T T G C C C G –

mutation 2:
– C A T C C A A A T T T T G C C C G –

mutation 3:
– C A T C C A A C T T G T T G C C C G –

c (i) Explain the effect of each of the mutations shown above. [6]
(ii) State, with a reason, which of the three mutations would have the least effect
on the primary sequence of the polypeptide formed. [1]
d Explain the consequence of gene mutation, with reference to the condition known
as phenylketonuria (PKU). [3]
UCLES A Modular Sciences: Central Concepts in Biology June 1998 Section A Q2

Chromosome mutation

14

Heritable changes in genetic material can occur at the level of the chromosomes as well as at that of individual genes. These **chromosome mutations** are gross changes in the genome: they involve the *number* or the *structure* of chromosomes. In eukaryotes, the term 'genome' refers to the sum total of genes in the basic set of chromosomes.

Change in chromosome number

There are two sources of variation in the number of chromosomes: euploidy and aneuploidy.

Euploidy

Euploidy is variation in the number of whole sets of chromosomes. We regard diploids, with two sets of chromosomes in their nuclei, as being the normal form in eukaryotes. Monoploids (or haploids) have only one set. They develop from unfertilised eggs and are very rare. Organisms with three or more complete sets of chromosomes are known as polyploids. Polyploids are widespread and we will consider them in detail. The first thing to understand is how to represent their chromosome numbers.

Polyploidy

Chromosome numbers in polyploids

The convention for writing chromosome numbers, as explained in Chapter 2, is as follows:

x = the *basic number* of different chromosomes in a haploid set
n = the number of sets of chromosomes in the gametes; that is the *gametic number*
$2n$ = the number of sets of chromosomes in the zygote; that is, the *zygotic number*.

In a diploid, of course, $x = n$, and we write the chromosome number as, for example, $2n = 2x = 20$ (for example, maize). The $2n$ indicates that we are giving the zygotic number, and the $2x$ tells us that it is a diploid with *two sets* of 10 chromosomes in its somatic cells. To denote the level of polyploidy we simply use the appropriate value of x:

$$
\left.
\begin{array}{ll}
2n = 2x = 14 & \text{diploid} \\
2n = 3x = 21 & \text{triploid} \\
2n = 4x = 28 & \text{tetraploid} \\
2n = 5x = 35 & \text{pentaploid} \\
2n = 6x = 42 & \text{hexaploid}
\end{array}
\right\} \text{polyploid series based on } x = 7
$$

We can only write the chromosome number in this precise way when we know that we are dealing with a polyploid, and when we are certain of its basic number. Quite often we don't know whether we are dealing with a polyploid or not, and then we give the zygotic number as (for example) $2n = 42$ and the gametic number as $n = 21$. The value of x is unknown.

Polyploidy is rare among animals. For this reason their chromosome numbers are often written without reference to x; for example, $2n = 46$ in humans. Diploidy is taken for granted.

Occurrence of polyploidy

Polyploidy, by and large, is restricted to the plant kingdom. It is very rare in animals. The reasons for this disparity are not really known. Complications with sex-determining chromosomes are a possibility; as is some kind of physiological disturbance. Where polyploids do arise in animals they usually fail to develop, or else they are aborted, as in mammals.

Among flowering plants (angiosperms), on the other hand, polyploidy is common. Of angiosperms, 35% are either polyploid species, or they are species with polyploid *races* as well as diploids. Most of what we have to say about polyploids therefore refers to plants. A well-known example is the cultivated potato, *Solanum tuberosum* ($2n = 4x = 48$). Its chromosomes are shown in Figure 14.1, together with those of a wild diploid relative.

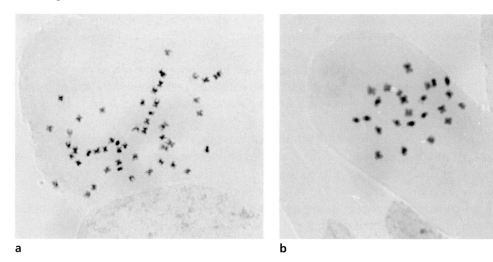

a   b

Figure 14.1 The cultivated potato, *Solanum tuberosum*, is a tetraploid with 48 chromosomes in its somatic cells ($2n = 4x = 48$) **(a)**. Its wild relative, *S. brevidens* **(b)** is a diploid ($2n = 2x = 24$). The basic number for the genus is $x = 12$. (Chromosomes photographed at c-mitosis in cells of the root meristem, ×2000)

Kinds of polyploids and their origin

There are basically two kinds of polyploids: autopolyploids and allopolyploids, and they arise in different ways.

Autopolyploids are polyploids with more than two sets of chromosomes from within a single species. They arise by spontaneous chromosome doubling. The most frequent way in which this happens is by spindle failure at meiosis, giving unreduced gametes (Figure 14.2, overleaf).

175

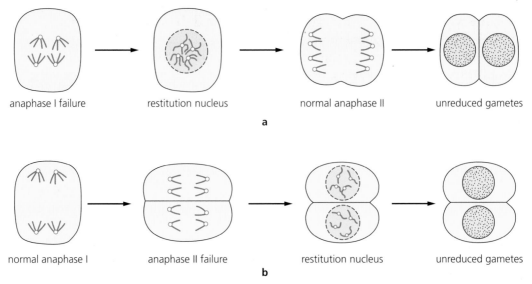

anaphase I failure   restitution nucleus   normal anaphase II   unreduced gametes

a

normal anaphase I   anaphase II failure   restitution nucleus   unreduced gametes

b

Figure 14.2 Unreduced diploid gametes can result from a failure in the first **(a)** or second **(b)** meiotic division. In **(a)** the chromosomes do not separate properly at anaphase I, but instead form a common nucleus, called a 'restitution nucleus'. This nucleus then undergoes the second division to form two unreduced diploid gametes. In **(b)** anaphase II is abnormal. The chromatids do not separate but form two restitution nuclei that again give two diploid gametes

When an unreduced diploid gamete ($2x$) fuses with a normal haploid one (x) a triploid ($3x$) is produced. The union of two unreduced $2x$ gametes gives an autotetraploid. Triploids can also arise from crossing between diploids and tetraploids.

Another source is spindle failure at mitosis: this gives direct doubling of the somatic chromosome number and leads to the production of a polyploid cell. Derivatives of the doubled up cell, be they $4x$, $6x$ or $8x$, may then form a polyploid sector or branch in an otherwise $2x$, $3x$ or $4x$ plant.

Autopolyploids have chromosome sets that are all homologous with one another. In the case of a tetraploid, with a basic number of $x = 3$, we may represent them as in Figure 14.3.

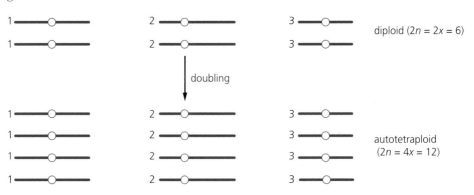

diploid ($2n = 2x = 6$)

doubling

autotetraploid ($2n = 4x = 12$)

Figure 14.3

Examples of autopolyploids are the celandine (*Ranunculus ficaria*, $2n = 4x = 32$, $2n = 5x = 40$ and $2n = 6x = 48$), the hyacinth (*Hyacinthus orientalis*, $2n = 3x = 24$ and $2n = 4x = 32$) and some varieties of apple, for example Cox's Orange Pippin (*Malus pumila*, $2n = 3x = 51$).

Allopolyploids are polyploids with multiple sets of chromosomes from more than one species. The simplest way in which they can arise is by doubling in a hybrid between two related diploids. Another way is simply by crossing between two species that are already polyploid – say two tetraploids – in which case there is no need for doubling. They may also derive from hybridisation involving more than two species (see Box 14.1). Allopolyploids are different from autopolyploids in that the chromosome sets contributed by the two (or more) parent species are not completely homologous (Figure 14.4).

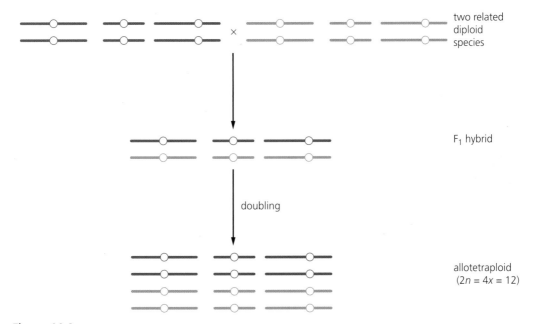

two related diploid species

F$_1$ hybrid

doubling

allotetraploid ($2n = 4x = 12$)

Figure 14.4

Examples of allopolyploids are to be found in the hempnettle, *Galeopsis tetrahit* ($2n = 4x = 32$, *G. pubescens* × *G. speciosa*), the tobacco, *Nicotiana tabacum* ($2n = 4x = 48$, *N. sylvestris* × *N. tomemtosiformis*) and in *Primula kewensis* ($2n = 4x = 36$, *P. floribunda* × *P. verticillata*).

Experimental production of polyploids

Chromosome doubling can be induced experimentally by the use of chemicals that interfere with spindle formation. Colchicine is the most widely used substance for this purpose. It disrupts the organisation of the spindle microtubules, but when it is withdrawn the cells can again divide normally. Plant breeders immerse young seedlings in a dilute aqueous solution of colchicine to make new polyploid varieties of crop plants.

Consequences and significance of polyploidy

Polyploids are important in nature because of the way in which they affect:

1 physiology and development
2 genetic variation
3 meiosis and reproduction.

For these reasons they differ in their adaptation from diploids and may be able to colonise and occupy new and different habitats to those of their diploid relatives. In the case of allopolyploids, entirely new species may be created.

Polyploids have larger nuclei than their diploid progenitors, and this in turn gives rise to an increase in cell size and an altered pattern of development and morphology. Leaves tend to be larger and thicker, the growth rate is slower, and there may be physiological differences in response to certain environmental factors (temperature, rainfall). These **gigas** effects are not invariable and they are generally much more pronounced in newly formed polyploids than in those that are well established.

The consequences in terms of genetic variation differ between autopolyploids and allopolyploids. In an autopolyploid, for instance, which has two allelic forms of a gene at a given locus (**A** and **a**), we can have five genotypes in the population – **AAAA, AAAa, AAaa, Aaaa, aaaa**. In a diploid, we have only three (**AA, Aa, aa**). One result is that the polyploid population will contain a much higher proportion of heterozygotes (**AAAa, AAaa, Aaaa**) and a greatly reduced fraction of homozygotes (**AAAA, aaaa**). We will not attempt to explain why this is so: suffice it to say that the pattern of genetic variation will be different from that in a diploid.

There are also complications in the pairing and distribution of chromosomes at meiosis. The point is that autopolyploids have problems in distributing their chromosomes in 'balanced' sets into the gametes. This is especially true in the odd-numbered ones ($3x$, $5x$, and so on). Their gametes are often inviable (that is, fail to develop) and the plants do not set seed properly. It is for this reason that autopolyploids have evolved the various methods of asexual reproduction (bulbs, corms, tubers, and so on). Even-numbered polyploids have much less of a problem.

Allopolyploids combine the genetic qualities of two different species into a new fertile hybrid. The F_1 from diploid parents is usually sterile because chromosomes from the two parent species are not homologous and cannot pair. After doubling, each chromosome has an identical partner. The chromosomes can then associate in pairs and the new species can behave as a diploid at meiosis.

The Russian geneticist Karpechenko (1928) first explained how chromosome doubling in a species hybrid could produce a fertile allopolyploid. The cross he made was between the radish (*Raphanus sativus*, $2n = 2x = 18$) and the cabbage (*Brassica oleracea*, $2n = 2x = 18$). The most important example of a natural allopolyploid is the bread wheat, *Triticum aestivum* ($2n = 6x = 42$), which was taken into cultivation by humans more than 6000 years ago (Box 14.1).

Box 14.1 Bread wheat – origin of an allohexaploid

The bread wheat, *Triticum aestivum*, is the most widely cultivated plant in the world. It is an allohexaploid, $2n = 6x = 42$ (Figure 14.6, overleaf). Three diploid wild grasses are involved in its origin, each one containing a different genome. The three genomes – that is, the three different basic sets of chromosomes – are represented as A, B and D. The source of the B genome is unknown. The A genome came from *T. monococcum*, which hybridised naturally with the B genome donor to give the wild allotetraploid *T. turgidum*. A cultivated variety of the tetraploid (var. *dicoccum*) crossed naturally with wild *T. tauschii* (growing as a weed) to give the cultivated allohexaploid, *T. aestivum*. Modern varieties of bread wheat have their origin in these natural hybridisation events, which took place more than 6000 years ago (Figure 14.5).

Figure 14.5 The origin of bread wheat

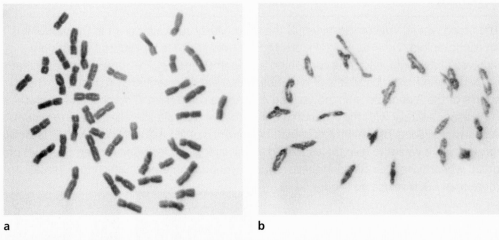

a   b

Figure 14.6 Chromosomes of bread wheat. C-mitosis in a root meristem cell **(a)**. At MI of meiosis the chromosomes within each of the three genomes come together in pairs to form 21 bivalents **(b)**. This ensures their perfect segregation and complete self-fertility

The distinction between autopolyploids and allopolyploids

We have described chromosome pairing behaviour in autopolyploids and allopolyploids in simple terms and have left out many of the details and complications. In reality, the difference between these two groups is not usually as clear-cut as we have suggested. The distinction may be blurred because the parent species that hybridise together to form allopolyploids vary widely in their degree of relatedness. In the case of *Raphanus sativus* × *Brassica oleracea*, the hybrid is **intergeneric** (that is, the two species are not even in the same genus), and the two chromosome sets in the F_1 are widely divergent in their structure and genetic organisation.

Hybrids between two closely related species from the same genus (for example, the ryegrasses *Lolium perenne* × *L. temulentum*) have chromosome sets with a much greater degree of **homology**. They can sometimes form bivalents at meiosis in the F_1, and a mixture of both bivalents and multivalents at the allopolyploid level. Their chromosome pairing is intermediate between that of an autopolyploid and an extreme form of allopolyploid.

Another complication arises from the fact that pairing behaviour can be controlled by single genes, as well as by structural divergence between different complements – which is the case in wheat.

Aneuploidy

Aneuploidy refers to change in number involving only part of a chromosome set. One or more whole chromosomes may be absent from, or in addition to, the diploid or polyploid complement. The absence of a chromosome in a diploid is denoted as $2n = 2x - 1$, and the addition by $2n = 2x + 1$.

Origin of aneuploidy

The irregular distribution of chromosomes at meiosis in polyploids, particularly the odd-numbered ones, is one source of aneuploidy. This applies mainly to plants, since polyploidy is rare in animals. Aneuploids may also be produced by a process of **non-disjunction**. This is a mistake in chromosome separation in which a pair of chromosomes (or chromatids) passes to the same pole of a cell instead of to opposite poles. It can happen at mitosis or meiosis. It is more likely to give aneuploid progeny when it occurs at meiosis (Figure 14.7). This is the way in which aneuploids arise in animals, including humans.

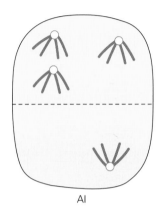 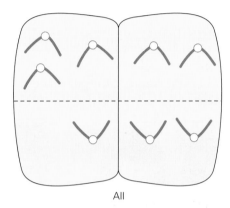

AI   AII

Figure 14.7 Diagram showing how aneuploidy arises by non-disjunction at AI or AII of meiosis. In the AI cell a pair of chromosomes is also shown disjoining normally. At AII both cells are shown – one is normal and the other has chromatid non-disjunction in one of its chromosomes

Consequences of aneuploidy

Aneuploidy causes an imbalance in the chromosome complement and nearly always results in an abnormal phenotype. A well-known example of aneuploidy in humans is shown in Figure 14.8, overleaf. In plants the effects are much less severe in polyploids than in diploids. The degree of abnormality depends on which *particular* chromosome is missing or additional, and in general an extra chromosome is less deleterious than a missing one.

Aneuploidy in humans

The most common aneuploids in humans are those involving the sex chromosomes. The imbalance results in abnormalities in development but is generally not lethal.

A well-known example is **Turner's syndrome** (XO), in which there is only one X and no Y. The affected individuals are female, and the condition occurs at a frequency of about 1 in 3000 female births. Although about 98% of fetuses with this condition spontaneously abort, those that do survive are often of almost normal phenotype. Some effects that may occur include a flap of 'webbed' skin on the neck and faulty spatial perception. In addition, affected individuals have underdeveloped secondary sexual characteristics and are sterile.

Aneuploidy in males includes **Klinefelter's syndrome** (XXY and XXXY), which occurs at a frequency of about 1 in a 1000 male births. Affected people have additional Xs but they develop as males due to the presence of the Y chromosome. They tend to have more feminine proportions, including larger breasts than usual, and they may be taller than average. Sometimes the condition is associated with mental retardation, but not always. Men with Klinefelter's syndrome are infertile, with about 5% of men who go to infertility clinics being found to have the syndrome. Some men live entirely unaware that they have the syndrome until they try to have children.

Another category that occurs in about 1 in 1000 births, is **XYY syndrome**. These males are usually fertile and may be normal in phenotype, although the condition is often associated with some degree of mental handicap. There appears to be a significantly higher proportion of XYY men among socially deviant males than there is within the population in general. However over 95% of XYY men lead a normal life and so it certainly cannot be said that an XYY genotype causes men to be criminals.

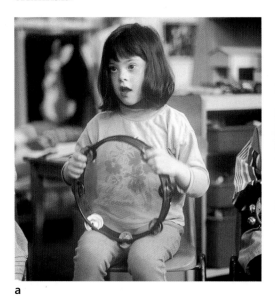

a

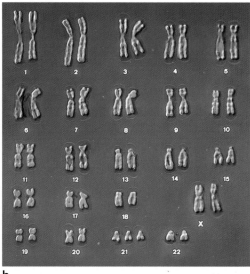

b

Figure 14.8 Young person with Down's syndrome **(a)**. Karyotype of a person with Down's syndrome due to trisomy of chromosome 21 **(b)**

Down's syndrome is the best known of the autosomal aneuploid conditions (Figure 14.8). One form is due to trisomy (three copies) of chromosome 21 (Figure 14.8b). The phenotype is characterised by mental retardation and certain distinctive physical features. These features can include malformed ears, speckles in the eyes, a broad flat nose and large tongue. People with Down's also tend to be shorter than average and have hypotonia, a relaxation of the muscles that may, for example, cause the mouth to droop open. Many Down's patients do not develop reading or writing skills, and have a limited vocabulary. Nevertheless, people with Down's tend to be sociable and good-natured, and improved provision and education opportunities in recent decades have dramatically enhanced their quality of life in many cases. The condition is also associated with an increased likelihood of heart malfunction, leukaemia, and impaired functioning of the immune system.

An important aspect of Down's syndrome is its incidence in relation to the age of the mother (Figure 14.9). Women over the age of 40 who become pregnant are advised to have the chromosomes of the fetus examined. This can be done because a few cells get sloughed off the fetus into the amniotic fluid. A sample of this fluid containing the cells is then removed through the wall of the uterus (with a syringe) and the cells are multiplied up in tissue culture for chromosome analysis. This method of sampling and analysis of the amniotic fluid is known as **amniocentesis**.

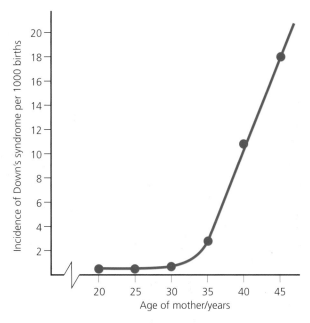

Figure 14.9 Graph showing the effect of maternal age on the incidence of children with Down's syndrome

Trisomy is also known for chromosomes 18 (Edward's syndrome) and 13 (Patau's syndrome), although sufferers of these conditions usually die within a few months of birth. Chromosomes 18 and 13 are bigger than 21, which is the smallest chromosome in the human complement. Aneuploidy for other larger chromosomes in the complement, aside from 21, 18, 13, X and Y, is lethal. It is estimated that out of every 1 000 000 conceptions in humans there are 150 000 spontaneous abortions, of which 75 000 are due to some kind of chromosome abnormality (aneuploidy, polyploidy and structural changes). Out of the remaining 850 000 live births some 833 000 survive and 5165 of these carry a chromosome mutation. The frequency of chromosome mutation in the human population is therefore of the order of 0.61% of live births; that is, more than 1 in 600.

Change in chromosome structure

Variation in chromosome structure results from breakage and from errors in crossing over. Breakages may, or may not, be followed by rejoining of the broken ends. As these changes are the result of random events, it is unlikely that both members of a pair of homologous chromosomes in diploid cells will be affected in the same way at the same time. Consequently, *structural* changes arise in heterozygous form. We recognise four main categories: deletions, duplications, inversions and translocations.

Deletions

A **deletion** (or **deficiency**) refers to the loss of a chromosome segment (Figure 14.10).

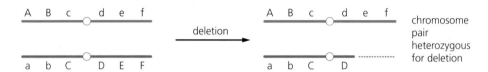

Figure 14.10

Deletions are usually lethal even in the heterozygous condition. As homozygotes they will certainly be lethal if the genes that are lost are concerned with some essential function. A well-known example of a heterozygous deletion in humans is the '*Cri-du-chat*' syndrome, which is due to the loss of a segment from chromosome 5. It causes mental retardation, abnormalities of the face and head, and a characteristic high-pitched cry, which resembles that of a cat in distress.

Duplications

A **duplication** occurs when a chromosome segment is present more than twice in a diploid (Figure 14.11).

Figure 14.11

Duplications can arise from errors in crossing over. They affect the phenotype because of the numbers of copies of the genes involved (dosage of genes). The effect depends upon the particular segment concerned, but clearly duplications are less harmful than deletions as there is no loss of genetic material. Duplications are important in evolution. When more than one copy of a gene is present, the redundant one is free to mutate and to evolve a new function. The α and β haemoglobin genes are thought to have arisen in this way.

Inversions

An **inversion** is the reversal of the gene order that may result when two breaks occur in the same chromosome (Figure 14.12).

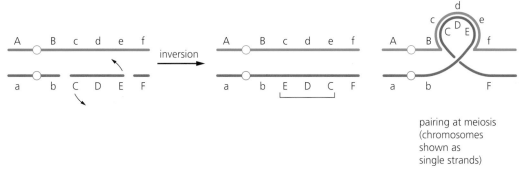

pairing at meiosis
(chromosomes
shown as
single strands)

Figure 14.12

There is no loss or duplication of genetic material in an inversion, but there is a problem at meiosis to do with pairing in the heterozygote. The two homologous chromosomes cannot align themselves side by side in the same way that they would normally do at prophase I, because the gene order is reversed in one of them. They overcome this difficulty by forming a 'reverse pairing loop' (Figure 14.12). If crossing over takes place within the loop, the chromosomes have difficulty in separating and half of the gametes that are produced are inviable. That is why inversions are important – they prevent recombination taking place between adaptive combinations of linked genes within the inverted region. Crossing over can take place elsewhere within the bivalent, of course. There is no difficulty at meiosis in the inversion homozygote (that is, where the same inversion is present in both homologous chromosomes), and no problem at mitosis in either heterozygotes or homozygotes.

Translocations

In a **translocation**, a segment is transferred from one chromosome to another, non-homologous one. If two chromosomes exchange segments, the translocation is then sometimes called an **interchange** (Figure 14.13).

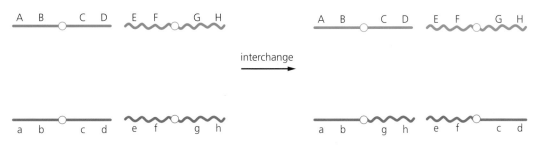

Figure 14.13

As with inversions, these translocations do not affect mitosis in any way, but they cause complications at meiosis as heterozygotes. The reason is that two pairs of chromosomes now have some homologous parts in common, and four chromosomes can associate together at prophase I. We will not attempt to explain the complexities of chromosome pairing and segregation in an interchange heterozygote. We will simply say that there may be infertility due to the production of inviable gametes, and that interchanges are important because they link together genes in different pairs of chromosomes and prevent their independent segregation. As with inversions, they therefore regulate the pattern of genetic variation in natural populations.

Cancer cells are usually associated with chromosome abnormalities, either as a cause or consequence of unregulated cell division. Some chromosome mutations are found consistently in certain types of cancer, especially in leukemias. In humans, chronic myelogenous leukaemia is frequently associated with a *reciprocal* translocation between chromosome 22 and chromosome 9 in blood stem cells. A part from one end of each chromosome is broken off and re-attached to the broken end of the other chromosome. The altered chromosome 22 is known as the 'Philadelphia chromosome' because it was first discovered in cancer patients in Philadelphia. The breakpoints in each chromosome occur within a gene, so that translocation results in the ends of the two different genes being physically joined together. The protein product of this 'fusion' gene somehow causes white blood cells to become cancerous.

Summary

◆ Chromosome mutations cause major changes in the quantity and quality of the genetic material in the nucleus.

◆ Numerical changes involving whole sets of chromosomes affect all aspects of development, genetic variation, meiosis and reproduction. They may even give rise to entirely new species, including several that are of use to humans.

◆ Changes in part of a set (aneuploidy) are generally deleterious because they upset the genic balance between different chromosomes within the complement. Aneuploidy is a major problem in the human race.

◆ Structural changes are important in terms of evolution of chromosomes, and for the way in which they can regulate the pattern of genetic variation in natural populations – although this aspect is not discussed in detail.

Further reading

1 C. D. Darlington (1956) *Chromosome Botany*. George Allen and Unwin.
2 G. L. Stebbins (1971) *Chromosome Evolution in Higher Plants*. Edward Arnold.
3 S. E. Antonarakis (1998) 'Ten years of *Genomics*, chromosome 21, and Down syndrome'. *Genomics* **51**(1): 1–16.
4 W. S. Clug and M. R. Cummings (2000) *Concepts of Genetics*. Sixth edition. Prentice Hall, Inc. Website:
   http://cw.prenhall.com/bookbind/pubbooks/klug3/
5 J. R. Lupski, J. R. Roth and G. M. Weinstock (1996) 'Chromosomal duplications in bacteria, fruit flies and humans'. *Am. J. Hum. Genet.* **58**: 21–26.
6 M. W. Strickberger (1996) *Evolution*. Second edition. Jones and Bartlett.
7 The Cytogenetics Gallery. Photos, karyotypes and brief details of the cytogenetics of various human genetic disorders. Also some information on the mouse:
   http://www.pathology-washington-edu:80/cytogallery.html
8 E. Therman and B. Susman (1995) *Human Chromosomes*. Third edition. Springer-Verlag.

Questions

1 Give an illustrated account of how mutations occur. Describe, using examples, how mutations can cause genetic defects in humans.

2 Describe the different types of gene and chromosome mutations. Discuss their relevance to the evolution of new species.

3 **a** What are genes?
  b Explain the terms *gene-linkage*, *gene mutation*, *chromosome mutation* and *polyploidy*.

4 **a** Give *two* symptoms of Down's syndrome.
  b What chromosome abnormalities are responsible for Down's syndrome, and how might they occur?
  c The unborn babies of women over 40 years old may be screened for Down's syndrome. Describe how this might be done, naming the technique you describe.

5 **a** Why do closely related species rarely form hybrids in nature?
  b Black mustard (*Brassica nigra*, 2n = 16) plants have been produced experimentally that have 32 chromosomes. Explain how this might be done. The plants produced were almost totally sterile. Explain why this was so.
  c Hybridisation between normal black mustard plants and turnips (*Brassica rapa*, 2n = 20) has produced brown mustard (*Brassica juncea*, 2n = 36). Explain how such hybrids might be produced and what chromosome changes have occurred during the formation of the hybrid. Why can the brown mustard produce viable seed?

6 Give an account of chromosome mutations and describe how they may contribute to genetic variation.   [10]
London AS/A Biology and Human Biology Module Test B/HB1 January 1998 Q8

7 The diagram shows the karyotype of a boy with Down's syndrome.

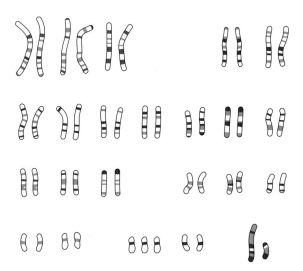

a State the number of chromosomes shown in this karyotype. [1]

b Describe the position of the Y chromosome on the karyotype (say which row it appears in, and what position in that row). [1]

c Down's syndrome is an example of polysomy, and may arise as a result of non-disjunction during the formation of a female gamete.

(i) State what is meant by the term *polysomy*. [1]

(ii) Explain how fertilisation by normal sperm of a female gamete formed by non-disjunction may result in Down's syndrome. [3]

London AS/A Biology and Human Biology Module Test B/HB1 June 1999 Q5

Genes in populations

So far we have largely been concerned with genetic material at the level of the cell and individual organism. We now consider how heredity and variation work at the level of populations and species (Chapters 15–17).

Our studies on inheritance have mainly been concerned with experimental crosses. We have always started off with matings between two pure-breeding parents, and have then followed the progenies through the F_1 and F_2 generations, or else through backcrosses.

At the level of the *population* the situation is different. In any particular generation, *all* of the adults in the population are potential parents, and with sexual outbreeding they can interbreed in all kinds of combinations. For a gene with two alleles, **A** and **a**, there are three possible genotypes, **AA**, **Aa** and **aa**, and these may occur in any proportions relative to one another. This means that, instead of dealing with defined ratios within families, we will be concerned with *frequencies* of the different genotypes in the population as a whole. The question that arises, therefore, is this: given certain genotype frequencies (say 36% **AA**, 48% **Aa** and 16% **aa**) in one generation, what will be their relative frequencies in the next generation, and in the generation after that? In other words, how are genes inherited in populations?

In this chapter, we will see that inheritance at the population level takes place by the same Mendelian processes of segregation and random combinations of pairs of alleles that we have studied earlier, and can easily be predicted by using a simple formula. The study of the genetic composition of populations is known as **population genetics**. The population geneticist tries to find out the frequencies of gene alleles, and genotypes, in natural populations and to study the factors that determine them. We will begin by defining and explaining some of the basic terms used in population genetics.

Populations, gene pools and allele frequencies

Populations

To the geneticist, the term 'population' has a precise meaning. A **population** is a local community of a sexually reproducing species in which the individuals share a common gene pool (Figure 15.1). We will mainly be concerned with populations of sexually reproducing organisms in which there is random mating (**panmixis**), with each member having an equal chance of mating with any other member of the population. Because the sharing of genes takes place essentially by Mendelian inheritance, such local communities are also referred to as **Mendelian populations** (or **demes**).

The largest exclusive group of potentially interbreeding individuals that can comprise a Mendelian population is the **species**, but it is rare for an entire species to form one random mating group. What we generally find is that the species is made up of a large number of local populations (demes) with varying degrees of 'gene flow' between them. At one extreme we may have an 'open population', which is subject to immigration of genes from other intercommunicating groups within the species, and at the other end of the range a 'closed population' with the only source of new alleles being mutation.

Generally speaking there are no clear-cut boundaries between one Mendelian population and another but, leaving this difficulty aside, the population geneticist looks on the Mendelian population as the basic unit of study and is interested to know how the genes are distributed and inherited within these populations, and how one local interbreeding group differs from another in its genetic composition.

The kind of population we have in mind might be a local cluster of wild garlic (*Allium ursinum*) in a wood, an isolated colony of the scarlet tiger moth (*Panaxia dominula*) in Southern England, a colony of house sparrows (*Passer domesticus*) inhabiting the buildings and hedgerows of an isolated village, a colony of black-headed gulls (*Larus ridibundus*) nesting in sand-dunes, or sticklebacks (*Gasterosteus aculeatus*) breeding in a pond.

Throughout the rest of this chapter, we will be concerned with sexual outbreeders, and not attempt to deal with any of the complications that arise from inbreeding and from various forms of **assortative** (non-random) **mating**.

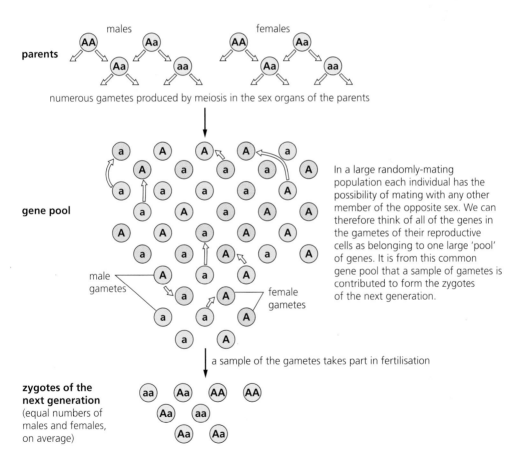

Figure 15.1 Diagram explaining the idea of a gene pool, using one gene with two alleles

Gene pools

In a Mendelian population, the sum total of the genes within the reproductive cells of all its members constitutes the **gene pool**. Reproduction in a population takes place, of course, between individuals of opposite sexes. But because mating is random, with each individual having an equal chance of mating with every other member of the population, we can think of all the genes in the gametes as belonging to one large 'pool' of genes (Figure 15.1). It is from this pool that a *sample* is taken into the zygotes of the next generation.

In thinking about the inheritance of genes in populations it is the gene pool, and the way in which its composition may or may not change over a number of generations, that must concern us – not what happens to any individual member. The unit of inheritance, and of evolutionary change, at the population level is the sample of the gene pool that is passed on from one generation to the next.

Allele frequencies

A gene with two alleles (**A** and **a**) will give rise in a population to three different genotypes (**AA, Aa, aa**); a gene with more than two alleles will give rise to correspondingly more. The relative proportions of the various genotypes present in a population form the **genotype frequency**. The **allele frequency** is the relative proportion of the alleles of a gene present in a population.

Allele frequencies and genotype frequencies are important because they are characteristics of a population that geneticists can estimate, and observe for change; and also because they describe the genetic composition of a population for a particular gene. In the study of population genetics, we rarely deal with more than one or two gene loci at the same time – otherwise the system is too complex.

In order to distinguish clearly between the terms 'genotype frequency' and 'allele frequency', and to see how we estimate their values, we will use the **MN** blood group alleles in humans as an example. The gene to which these alleles belong gives rise to antigenic proteins on the surface of the red blood cells during their formation. There are two forms of the antigen, one determined by the **M** allele and the other by **N**. Homozygotes (**MM, NN**) carry one form or the other and the heterozygotes (**MN**) produce both. The alleles are therefore *co-dominant* and each of the three genotypes can be easily distinguished and classified.

In a sample of 730 Australian Aborigines the numbers of individuals with the various blood group types were found to be as in Table 15.1.

Table 15.1 Blood group types among a sample of Australian Aborigines

Genotype	MM	MN	NN
Number of individuals	22	220	488
Genotype frequencies (%)	3.0	30.1	66.9

The genotype frequency is the percentage of each genotype in the total sample; for example, the genotype frequency for **MM** in Table 15.1 is $22/730 \times 100 = 3.0\%$.

In the same way that we think of each individual as being representative of one genotype, so we think of each diploid individual within a population as being represented by one pair of alleles for the purpose of calculating allele frequencies. (We have to overlook the fact that each individual is made up of many millions of cells.)

To calculate allele frequencies, using numbers of genotypes, the procedure is as shown in Table 15.2.

Table 15.2

Genotype	MM		MN		NN		
Number of genotypes	22		220		488		
Number of M alleles	44	+	220				
Number of N alleles			220	+	976		
Total number of alleles	44	+	440	+	976	=	1460

$$\text{Frequency of allele } \mathbf{M} = \frac{44 + 220}{1460} \times 100 = 18\%$$

$$\text{Frequency of allele } \mathbf{N} = \frac{976 + 220}{1460} \times 100 = 82\%$$

It is usual in population genetics to express both allele frequencies and genotype frequencies as decimal fractions, rather than percentages, so that the arithmetic can be handled more easily. The estimates of allele frequencies for the Australian Aborigines are therefore:

$$\text{Freq. } \mathbf{M} = 0.18$$
$$\text{Freq. } \mathbf{N} = 0.82$$

Since we are working with fractions, and we have only two alleles, then:

$$(\mathbf{M} + \mathbf{N}) = (0.18 + 0.82) = 1.0$$

To calculate allele frequencies using frequencies of genotypes, instead of numbers, the procedure is even simpler: it is based on the reasoning that all of the alleles in **MM** genotypes and half those in **MN** are **M**, and the same for **N** (Table 15.3).

Table 15.3

Genotype	MM		MN		NN		
Frequency (as decimal fraction)	0.03		0.30		0.67		
Frequency of allele M	0.03	+	$\frac{1}{2}$ 0.30			=	0.18
Frequency of allele N			$\frac{1}{2}$ 0.30	+	0.67	=	0.82

That is:

$$\text{Freq. } \mathbf{M} = \mathbf{MM} + \tfrac{1}{2}\mathbf{MN} = 0.18$$
$$\text{Freq. } \mathbf{N} = \mathbf{NN} + \tfrac{1}{2}\mathbf{MN} = 0.82$$

Constant allele frequencies

The way in which genes are inherited in populations is no different in principle from the way in which they are transmitted in simple experimental crosses. All we require, at the population level, to predict the outcome of random mating among a mixture of genotypes is a simple extension of Mendel's idea of the segregation and chance combination of pairs of alleles that were described in Chapter 3. Provided that we know the frequencies of the two alleles (**A** and **a**) in the population, and provided that the three genotypes (**AA**, **Aa**, **aa**) all have equal chances of contributing gametes, then we can say precisely what frequencies we will have of **AA**, **Aa** and **aa** in the following, and in all subsequent generations.

Suppose, for example, that we have two alleles of a gene, **A** and **a**, present in a sample of gametes in frequencies of $0.6\,\mathbf{A} + 0.4\,\mathbf{a} = 1.0$. At fertilisation, the random combinations of these gametes will give genotypes among the progeny as shown in Figure 15.2. These genotypes will become the parents of the next generation and they in turn will contribute their gametes in frequencies of:

$$\mathbf{A} = \mathbf{AA} + \tfrac{1}{2}\mathbf{Aa} = 0.36 + (\tfrac{1}{2} \times 0.48) = 0.6$$
$$\mathbf{a} = \mathbf{aa} + \tfrac{1}{2}\mathbf{Aa} = 0.16 + (\tfrac{1}{2} \times 0.48) = 0.4$$

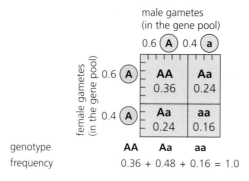

Figure 15.2

The allele frequencies in the gametes have therefore not changed between one generation and the next. The genotype frequencies will likewise be the same as they were before, and if we repeat the cycle over again we will find that both gene and genotype frequencies will remain constant *generation after generation*, indefinitely – provided that there are no disturbing influences (as described in Chapter 16) to prevent the different genotypes from making equal contributions to the progeny.

The reason for constant gene and genotype frequencies is the *binary* nature of inheritance. At fertilisation, alleles combine randomly in pairs in the diploid zygotes, and when they are present in frequencies of $0.6\,\mathbf{A} + 0.4\,\mathbf{a}$ they will always give genotype frequencies of $0.36\,\mathbf{AA} + 0.48\,\mathbf{Aa} + 0.16\,\mathbf{aa}$, simply because these are the chance combinations for alleles in these frequencies. During meiosis, the opposite process takes place and segregation, or separation, of the pairs of alleles *undoes* these chance combinations and releases the alleles back into the gametes in their original proportions (Figure 15.3).

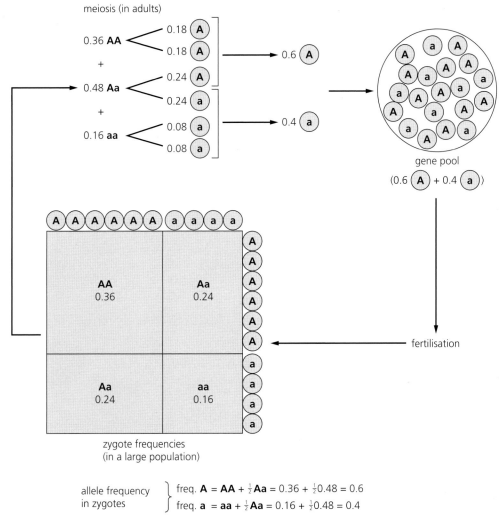

Figure 15.3 allele frequency in zygotes equations:

freq. **A** = **AA** + ½**Aa** = 0.36 + ½0.48 = 0.6

freq. **a** = **aa** + ½**Aa** = 0.16 + ½0.48 = 0.4

Figure 15.3 The inheritance of genes in populations takes place by the same processes of heredity as in simple Mendelian crosses. These processes involve the segregation, or separation, of pairs of alleles at meiosis, followed by their random combinations in pairs in the zygotes during fertilisation

The Hardy–Weinberg law

The English mathematician G. H. Hardy and the German geneticist W. Weinberg discovered this rule of constant gene and genotype frequencies independently of one another in 1908. They showed that it could be used as a general law whatever the allele frequencies.

To see how this law works, we will use the standard convention where p represents any given frequency of allele **A** and q (that is, $1 - p$) the corresponding value for **a**. Allele frequencies in a sample of gametes are $p\mathbf{A} + q\mathbf{a} = 1.0$. In our example given above $p\mathbf{A}$ was 0.6 and $q\mathbf{a}$ was 0.4. The random combinations of these gametes give the genotype frequencies shown in Figure 15.4 (overleaf).

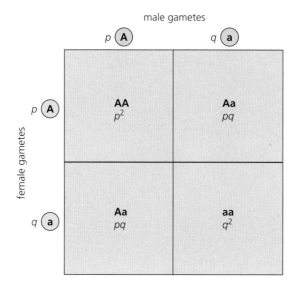

Figure 15.4

This combining process may be represented more conveniently in the binomial expression:

$$(pA + qa)(pA + qa) = p^2AA + 2pqAa + q^2aa$$

Or simply:

$$(p + q)^2 = p^2 + 2pq + q^2$$

These genotypes become the parents of the next generation and we can calculate the allele frequencies in their gametes, in terms of p and q, as follows:

$$\text{Freq. allele } \mathbf{A} = \mathbf{AA} + \tfrac{1}{2}\mathbf{Aa}$$

$$= p^2 + \frac{2pq}{2}$$

$$= p^2 + p(1 - p), \text{ because } q = 1 - p$$

$$= p^2 + p - p^2$$

$$= p$$

$$\text{Freq. allele } \mathbf{a} = \mathbf{aa} + \tfrac{1}{2}\mathbf{Aa}$$

$$= q^2 + \frac{2pq}{2}$$

$$= q^2 + q(1 - q)$$

$$= q^2 + q - q^2$$

$$= q$$

The allele frequencies of $A = p$ and $a = q$ are therefore constant between one generation and the next; and the genotype frequencies will also remain unchanged, for the reasons given earlier (see page 194).

This relationship between gene and genotype frequencies is known as the **Hardy–Weinberg law** or (**principle**) and it is the basic rule that forms the foundation of all population genetics. The Hardy–Weinberg law states that in a large randomly mating population there is a fixed relationship between gene and genotype frequencies and – in the absence of selection, migration and mutation – these frequencies remain constant from generation to generation.

The equation $(p + q)^2 = p^2 + 2pq + q^2$, which gives the relationship between genes and genotype frequencies, is known as the Hardy–Weinberg formula.

Given the qualifying conditions mentioned above, we will expect the three genotypes **AA**, **Aa** and **aa** to be present in the same equilibrium proportions of $p^2 : 2pq : q^2$ in every generation, for whatever values of p and q we are dealing with, and to remain in these fixed frequencies for as long as the conditions needed for stability prevail. Under these circumstances a **genetic equilibrium** will exist and the gene and genotype frequencies will be in accordance with the Hardy–Weinberg formula.

It also follows from what has been said above that only a single generation of random mating is required to establish a genetic equilibrium between gene and genotype frequencies. This is because genotype frequencies are determined only by the allele frequencies of the gametes that give rise to them. Any sample of gametes, with any given allele frequencies – irrespective of where they come from – will always give progeny genotypes in the proportions $p^2 : 2pq : q^2$, because these are the chance frequencies coming from randomly combining pairs.

In Chapter 16, we will discuss how allele frequencies can change, and how the *equilibrium* between gene and genotype frequencies can be influenced by the forces of mutation, migration, genetic drift and natural selection.

Applications of the Hardy–Weinberg formula

The Hardy–Weinberg formula can be used by the population geneticist to study the genetic composition of populations and to test genotype frequencies to see whether or not they conform with equilibrium distributions. This can be done in situations where all three genotypic classes can be identified, as in the **MN** blood group genotypes. The allele frequencies are first found (as described earlier) and are used to calculate the expected distributions according to the Hardy–Weinberg formula; these expected values are then compared with the observed frequencies of the population sample using the chi-squared test (see page 53). When genotype frequencies do *not* conform to equilibrium distributions, we may suspect that some disturbing influences are at work, such as mutation, migration, genetic drift or natural selection (Chapter 16).

A second application of the law is in finding the allele frequencies, and distribution of genotypes, in cases of dominance where we are unable to distinguish between the homozygous dominants and the heterozygotes on the basis of phenotype. Such an example is provided by the gene in humans that controls the ability to taste the chemical phenylthiocarbamide (PTC). Tasting ability is conferred by the dominant allele **T**, so that both the homozygous dominant class (**TT**) and the heterozygotes (**Tt**) can taste this bitter substance, and the non-tasters are the double recessives (**tt**). If a large class of 200 students is tested, it may well happen that 130 (65%) are tasters and 70 (35%) are non-tasters (Table 15.4).

Table 15.4

Phenotype	Genotype	Frequency
tasters	TT + Tt	0.65
non-tasters	tt	0.35

We will follow the convention of representing the frequency of the dominant allele (**T**) by p and the recessive allele (**t**) by q; and $p + q = 1.0$. Making the *assumption* that the population is at equilibrium, and that we have none of the disturbing influences referred to above, the genotypes will occur in the following proportions according to the Hardy–Weinberg formula:

$$\begin{array}{ccccc} \text{TT} & & \text{Tt} & & \text{tt} \\ p^2 & + & 2pq & + & q^2 \end{array}$$

We know the value of q^2 to be 0.35, and by taking the square root we obtain the value of $q = 0.59$. Since $p + q = 1.0$, $p = (1 - q) = 0.41$. Substituting these values into the formula we can compare the genotype frequencies:

$$\begin{array}{ccc} \text{TT} & \text{Tt} & \text{tt} \\ 0.41^2 & 2(0.41 \times 0.59) & 0.59^2 \\ 0.17 & 0.48 & 0.35 \end{array}$$

or

$$\begin{array}{ccc} 17\% & 48\% & 35\% \end{array}$$

A third use of the law is in making 'models' in order to see how gene and genotype frequencies change in response to certain aspects of selection. The use of the Hardy–Weinberg law can also be extended to cover sex-linkage, and the more complex genetics involved in working with multiple alleles, with two or more loci at the same time, and with polyploidy.

Summary

◆ The geneticist looks upon a population as a large group of interbreeding individuals sharing a common gene pool. The unit of inheritance is the sample of this gene pool that is passed on from one generation to the next.

◆ Population geneticists try to estimate gene and genotype frequencies in order to study the genetic composition of populations and to find out how this sample of genes is inherited.

◆ It turns out that inheritance at the population level takes place by the simple Mendelian processes of segregation and random combination of pairs of alleles and can easily be predicted. The basic rule that is used for these predictions, and which forms the foundation of all population genetics, is the Hardy–Weinberg law.

◆ When genotypes are undisturbed (by mutations or selection), and can make equal contributions to their progeny, the Hardy–Weinberg law predicts constant allele frequencies.

Further reading

1 *Genes in Populations*. A free software program from UC Davis:
   http://animalscience.ucdavis.edu/extension/Gene.htm
2 *Genes in Populations*. A software program written for the Cornell Laboratory for Ecological and Evolutionary Genetics:
   http://www.bioweb.uncc.edu/faculty/leamy/popgen/geneinp.htm
3 D. L. Hartl (1988). *A primer of population genetics*. Sinauer Associates, Inc.
4 *Conserving Species, Populations, and Genetic Diversity*. For those interested in the application of population genetics:
   http://www.wri.org/biodiv/gbs-ix.html

Questions

1 **a** Explain what is meant by the term *population*, in the context of genetics.
 b What is meant by the terms *allele frequency* and *genotype frequency*?
 c State, in words, the Hardy–Weinberg principle.
 d Explain why gene and genotype frequencies remain constant from generation to generation in a population that is in Hardy–Weinberg equilibrium.
 e Write down the Hardy–Weinberg formula using p to represent the frequency of allele **A** and q the frequency of allele **a**.
 f Describe briefly, using examples, some of the ways a geneticist might use the Hardy–Weinberg formula to study populations.

2 In a randomly breeding population of mice, agouti coat (**A**) is dominant to non-agouti (**a**). Of a sample of mice taken from the population, 16% had non-agouti coats.
 a What are the frequencies of the agouti and non-agouti alleles in the population?
 b What proportion of the population would be expected to be homozygous for **A** and what proportion heterozygous?
 c If the population continues to breed randomly, what is the distribution of alleles in the next generation?

3 A farmer bought 800 lambs from a randomly breeding population of sheep. As the lambs grew, he discovered that 200 of them had an economically undesirable feature called crinkly-hair, caused by the recessive gene **cr**.
 a What is the frequency of the **cr** allele in the flock?
 b What proportion of the flock is likely to be heterozygous?
 c The owner removes all the crinkly-haired sheep from the flock and sends them to market. He allows the rest to breed freely. What proportion of the next generation of lambs would be expected to show crinkly-hair?

4 Thalassaemia major is a severe anaemia that is usually fatal in childhood. It is relatively common in Mediterranean populations. Thalassaemia minor is a mild anaemia, often difficult to detect at all.
 a Among people of Sicilian or Southern Italian ancestry now living in Rochester, New York, thalassaemia major occurs at a frequency of about one birth in 2400 while thalassaemia minor occurs at a frequency of about one birth in 25. Extrapolating these frequencies to a population of 10000, the distribution is approximately as show below.

Phenotype	Genotype	Frequency
thalassaemia major	**Th Th**	4
thalassaemia minor	**Th th**	400
normal	**th th**	95996

 b What are the frequencies of the **Th** and **th** alleles in the population? Show your reasoning.
 c Does the population approximate to the binomial distribution of genotypes expected from this allele frequency?

Natural selection and speciation

The genetic variation that exists in natural populations provides the basis for change. Gene frequencies are not always constant in the way predicted by the Hardy–Weinberg law (Chapter 15). Various forces act upon the variation and may alter the proportions of genotypes that are present in successive generations. The gradual change in the genetic composition of a population over a number of generations is known as **evolution** (organic evolution). In this chapter, we discuss the genetic basis of evolution and the way in which evolution can give rise to new species of living organisms.

Evolution

It could be argued that the theory of evolution of species by natural selection, which was formally proposed in the Darwin–Wallace lecture to the Linnean Society in 1858, is the most important idea ever put forward in the field of biology.

Box 16.1 Lamarck (1744–1829)

Lamarck subscribed to the view that species are not immutable, but could gradually change over long periods of time and evolve into new forms. According to his theory (1809), the modifications that an organism acquired during its lifetime in response to the environment could somehow be transmitted to its offspring. He believed that an organism's 'desire for improvement' was the driving force in evolution. The great size of the pectoral muscle of birds, for instance, he thought was due to the constant effort of straining to lift the bird into the air. Birds would improve their muscle structure, through the exercise of flying, and these improvements would then be passed on to their offspring. Likewise, organs for which there was little use would gradually diminish and become smaller in size and complexity.

Lamarck collected many examples to support his theory and his idea made a great impact at the time. It was important because it involved the notion of 'evolution', and was quite at variance with the thinking of most of his colleagues who firmly believed that species arose by individual acts of creation and were immutable.

Lamarck's theory foundered because it was based on an idea of heredity that was wrong. He believed in the inheritance of acquired characters: '... *all that has been acquired by their progenitors during their life is transmitted to new individuals* ...'. We now know that acquired modifications in the phenotype, as a result of use or dis-use of organs, cannot be inherited. Variations can only be transmitted if they are due to genetic differences. Lamarck had no knowledge of genetics and was unable to distinguish between genotype and phenotype, and thus between heritable and non-heritable variation.

Until the Darwin–Wallace lecture, there were two conflicting schools of thought about where species came from. One belief was that all species were individually created and remain constant in their form. The other belief was that they could evolve, and that new species could gradually arise from pre-existing ones. This latter belief had been held by several eminent biologists including Erasmus Darwin, the grandfather of Charles Darwin, and the French biologist Lamarck (Box 16.1). The importance of the Darwin–Wallace lecture was that it offered an explanation of how evolution could take place – through the gradual process of change brought about by natural selection. This idea caught the imagination of the majority of biologists and is the basis of the theory of evolution that we accept today.

Darwin and the theory of evolution by natural selection

Charles Darwin (1809–1882) devoted his whole life to the study of natural history and to a detailed series of observations and experiments in order to understand the adaptation of organisms, and their evolution.

An important phase of his work began at the age of 22, when he accepted the position of naturalist on Captain Fitzroy's survey ship H.M.S. Beagle, which the British navy despatched on a five-year voyage around the world (1831–1836). On this voyage Darwin encountered the rich fossil beds of South America where he discovered many species of extinct animals and noticed the close resemblance in design between the fossil forms and the living species of, for instance, armadillos, tapirs and anteaters.

He was greatly impressed too with the *variety* that he found within and between the different species of plants and animals. On the Galapagos Islands, off the coast of Ecuador, he noticed how different the principal groups of plants and animals were from those on the mainland. He observed too how the giant tortoises varied from island to island and that distinctive *races* could easily be recognised by the form and pattern of their shells. Also important were the 13 species of Galapagos finches that displayed remarkable adaptations suiting them to the different ecological niches they occupied (see page 217).

In 1838, two years after returning from the voyage of H.M.S. Beagle, Darwin read an essay by the English clergyman Thomas Malthus who wrote about the growth of human populations. Malthus pointed out that unless population growth was checked in some way, by disease, war, famine or birth control, the number of people on Earth would quickly increase until there was 'standing room only'. Darwin made calculations to show the same geometric increase was true of any species. He had already begun to formulate his ideas about the origin of species by evolution when Malthus' essay gave him the idea of 'natural selection'. He could see that all organisms produce an excess of offspring and if they are all able to survive then their population would increase in size *exponentially* (2, 4, 8, 16, 32, and so on) In fact, this does not happen, and numbers remain fairly constant from one generation to the next – except during the colonisation of new territory. He reasoned that the numbers are kept in check by competition for natural resources, such as space, food and mates. This competition he called the 'struggle for existence'.

The variation that occurs between individuals of a species, such as minute differences in structure, behaviour and so on, affects their chances of survival and reproduction. Darwin argued that individuals with variations that were advantageous, and which gave better adaptation to their environment, would be more likely to survive the struggle for existence, and to pass these favourable characteristics on to their offspring: '... *this preservation of favourable individual differences and variation, and the destruction of those that are injurious, I have called 'Natural Selection'.*' Spencer's words, 'survival of the fittest', were later adopted for this part of the theory.

Darwin realised that, in a given environment, a species will gradually accumulate variations that are most suited to that environment. If circumstances change, a different set of variations becomes advantageous and will replace the previous, less-well-adapted forms. He knew also that to be selected the variations must be heritable: '... *natural selection acts only by the preservation and accumulation of small inherited modifications* ...'.

Darwin worked for many years on his theory of evolution, corresponding with other naturalists and amassing a wealth of data to support his ideas. To back his case for '... *descent with modifications* ...' he drew on fossil evidence, comparative anatomy and embryology, and on the great diversity of form shown within domesticated species of plants and animals (Chapter 17). Twenty years after reading Malthus' book, he received a manuscript from another British naturalist, Alfred Russel Wallace, who had independently arrived at the same conclusion about the mechanism of evolution. Wallace graciously allowed papers by himself and Darwin to be read to the same meeting of the Linnean Society in 1858. The following year Darwin published his theory in a more extended form in his famous book *On the Origin of Species by Means of Natural Selection*. His work forms one of the corner-stones of biology.

Since Darwin's time a great deal more evidence has been accumulated to support the concept of organic evolution, which most biologists now accept as a reality. A brief summary of this evidence is given in Box 16.2, overleaf. What is not so well understood, but remains a matter of much lively debate, is *how* evolution takes place. Natural selection is generally accepted to be the main agent of change, but as explained later it is not the only one.

Figure 16.1 Charles Darwin

Box 16.2 Evidence for organic evolution

Fossil record

Fossils are the organic traces or impressions of once-living organisms buried by natural processes and subsequently permanently preserved. They show that living organisms did not all appear at once. The first forms were aquatic; terrestrial groups appeared much later. Some fossils show similarities with other fossils and also with living creatures, indicating a common ancestry. In general, the larger the age difference between a fossil and its living relative the greater the degree of divergence between them. This supports the idea of change by the gradual accumulation of many small differences over long periods of time. In some sedimentary rocks that are exposed at the edge of a cliff, or in a gorge, sequences of fossils can be found in strata of known ages, and the palaeontologist can show a series of gradual changes in the form of some organisms.

Some evolutionary histories, such as that of the horse (*Equus*), have been reconstructed in great detail – the record suggests that the horse evolved from a small dog-sized creature (*Eohippus*, the dawn horse), with five toes on the forelegs and four on the hindlegs, to the one-toed large animals of today.

Certain link organisms have been found that have structures intermediate between recognised major groups. A well-known example is *Archaeopteryx*, a reptile-like animal with feathers, representing the transition from reptiles to birds.

Comparative anatomy of living organisms

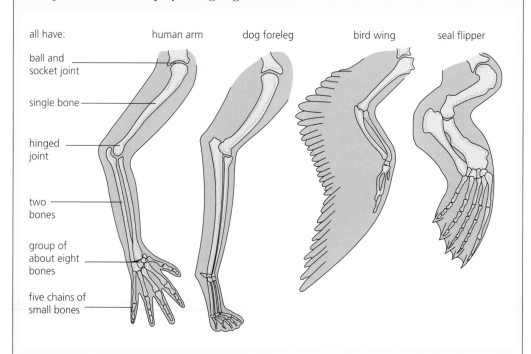

Figure 16.2

Similarities in the basic structure of different groups of plants and animals indicate their evolutionary relatedness. Furthermore, the comparison of *homologous* structures, which are parts of an organism that have similar structure but different function, supports the view that a basic structure can become modified or adapted to suit different needs. The best known example is the pentadactyl limb of vertebrates (Figure 16.2).

Homologous structures should not be confused with *analogous* structures, which are parts with similar function but different structure; for example, a bird's wing and an insect's wing.

Embryology

Resemblance between embryos is usually much greater than between adult forms. Embryos of vertebrates are almost all identical in early development (for example, they all form gill pouches) and evolutionists argue that this reflects their close affinity and descent from a common ancestral form.

Comparative biochemistry

All living organisms (except for some viruses) have DNA as their genetic material. All share the same universal genetic code and use their genetic information in the same way to code for proteins and RNAs that are virtually common throughout the plant and animal kingdoms. Many basic physiological processes, such as respiration, are the same in all species. The sequence of amino acids in certain proteins, such as cytochrome c (respiratory pigment in all cells), reveals remarkable patterns of molecular evolution. The number of amino acids by which various species differ in their cytochrome c can be used to construct a tree of their evolutionary relationships. Humans and chimpanzees share an identical sequence of 112 amino acids. These sequencing studies suggest a common origin for the cytochrome c gene in all present day species.

Geographical distribution

The world-wide distribution pattern of animals, and of plants, can be related to the past history of continental masses. The most distinctive faunas, for example, are found in Australia, which is the most isolated continent: it separated from mainland Asia before the placental mammals appeared. The marsupials have been able to thrive there, presumably because of the absence of the more successful placentals.

Direct evidence

Selection imposed by humans has brought about profound changes in the forms of our cultivated crops and domestic breeds of animals (Chapter 17). Changes in the genetic composition of populations, by natural selection and by human activities (Chapter 17), have demonstrated evolution in action.

Neo-Darwinism

Darwin's theory of evolution by natural selection had two major flaws:

1 Darwin had no idea about the *mechanism* of heredity
2 he was never able to demonstrate the *process* of natural selection actually taking place.

Darwin could not explain how variation arose or how it could be maintained over successive generations. This knowledge only became available after the rediscovery of Mendel's work and the development of the science of heredity. The advent of population genetics, which came later, made it possible to analyse changes in the genetic composition of populations and to demonstrate the forces responsible for evolutionary change.

The re-examination of Darwin's theory in the light of discoveries about the physical basis of inheritance (genes and chromosomes) is known as **neo-Darwinism** (neo = new).

Sources of variation

Darwin believed in blending inheritance and in various other ideas about how characters were transmitted, which turned out to be quite erroneous. As it happened, his basic theory did not depend on knowing the mechanism of heredity, although he was criticised for being unable to explain the origin of his 'spontaneous variation'.

We now know that the origin of genetic variation is due to the mutation of genes and chromosomes (Chapters 13 and 14). We know too that much of the variation we see in a population of outbreeders is due to recombination. Because of the particulate nature of heredity, alleles that combine together to form a heterozygote (**Aa**) retain their identity and can segregate out again unchanged (as **A** and **a**) in the gametes following meiosis. They can then *recombine* with the alleles of other genes to give an infinite variety of genotypes in the next generation (Chapter 6). This process of segregation and recombination can go on indefinitely as long as there is a supply of alleles available in the population.

We have seen earlier that all genes mutate at a constant and characteristic low frequency, about once per 100 000 gametes. The alleles that result are usually recessive and deleterious (Chapter 13). Because their frequency is so low, they may persist for several generations before they are eventually exposed and eliminated as homozygotes; that is, when two heterozygotes mate (**Aa** × **Aa**) some recessive homozygotes (**aa**) will appear among the progeny and will be selected against.

The elimination of segregating individuals that are homozygous for deleterious recessives, or which carry deleterious dominant alleles, is known as **genetic death**. It may take the form of mortality (for example, albino plants), sterility, failure to find a mate or any other circumstance that leads to reduced reproductive capacity. Natural populations are therefore imperfect and they carry a burden, or load, of hidden genetic defects. In every generation, a certain proportion of these recessives will segregate out as homozygotes and their presence will reduce the average level of *fitness* of the population. The extent to which the burden of deleterious alleles causes the population to depart from its optimum fitness is called its **genetic load**.

The genetic variability that is present in natural populations is the source of material upon which evolutionary forces act. Alleles and gene combinations that are unfavourable to an organism in one environment may well be advantageous in another one. Variation therefore gives a population the potential to change and to evolve in response to its environment. **Adaptation** is the term used to describe the structural or functional features that have evolved in an organism and enable it to cope better with its environment.

Natural selection: the principal force of change

Various forces operate upon the genetic variation present in natural populations to change their genetic composition and bring about adaptation and evolution. The principal force is natural selection, and this is the one to which we will give most of our attention.

Natural selection, in the way that we now see it, is defined as a process that determines gene frequencies in populations through unequal rates of reproduction of different genotypes. It comes about because the genotypes that comprise an outbreeding population vary in many aspects of their structure, physiology and behaviour, and some are better adapted to their environment than others. Where there is some impairment to the functioning of an individual, in its particular environment, then it will leave fewer than the average number of offspring and will contribute correspondingly fewer of its genes to the next generation. The gene, or genes, that cause the impairment will be selected against and may even be eliminated from the population. Those that are lethal will obviously disappear much more quickly than those with less harmful effects, because the selection against them will be that much stronger. Equally, genes that improve the fitness of individuals will improve the organism's chances of survival and reproduction and will therefore increase in frequency. The fitness value of a particular genotype will depend upon the environment in which it finds itself, as we will see below.

There are thus three principles to natural selection that we need to bear in mind.

1 There is heritable variation between the individuals of a natural population.
2 These individuals have unequal chances of leaving progeny.
3 These chances depend upon the conditions of the environment.

The best known examples of natural selection are those involving major gene **polymorphisms**, where there are two or more distinct forms of a species found in the same locality at the same time and in such frequencies (>1%) that the presence of the rarest form cannot be explained by recurrent mutation. To begin to understand how selection works we have to study genetic variation of this kind that is controlled by only one or two major gene loci.

The peppered moth

One of the most straightforward and widely known cases of a change in gene frequencies by natural selection is that in the peppered moth, *Biston betularia*, in Britain. It is an excellent example of a **transient polymorphism** (that is, the gradual replacement of one allele by another one).

The polymorphism involves a single gene locus with two main alleles. A recessive allele (**c**) determines the so-called 'typical' phenotype expressed in homozygotes (**cc**). These moths are light-coloured with a 'peppering' of black spots on the wings and body (Figure 16.3) and were the predominant form present in the first part of the 1800s. In 1849, a dark-coloured melanic variety called 'carbonaria' was reported for the first time in Manchester. The 'carbonaria' phenotype is due to a dominant mutation, **C** (**CC, Cc**). The two alleles are inherited in the predictable Mendelian manner in experimental laboratory matings, with **C** showing full dominance over **c**.

a
b

Figure 16.3 'Typical' and 'carbonaria' forms of the peppered moth (*Biston betularia*) at rest on trunks. A lichen-covered tree in an unpolluted area **(a)** and a lichen-free, soot-covered trunk in an industrial region **(b)**

During the second part of the nineteenth century, the dominant allele increased rapidly in frequency and by 1895 the 'carbonaria' variety comprised over 95% of the population in the Manchester area. The causes of this rapid alteration in gene frequency that Philip Sheppard once referred to as '*the most spectacular evolutionary change ever witnessed and recorded by man*', were extensively studied by Bernard Kettlewell in the 1950s. Kettlewell's work showed that the spread of the 'carbonaria' form was associated with the Industrial Revolution and the parallel spread of pollution caused by the fall-out from factory chimneys. The carbon particles and the SO_2 from atmospheric pollution killed the lichen on tree trunks and rocks and gradually changed the habitat of the moth (in the affected regions) from light-coloured, lichen-covered surfaces to areas that were bare and blackened with soot. Kettlewell showed a definite link between the pattern of spread of the 'carbonaria' and the presence of urban industrial pollution. The highest frequencies of the melanic form were in the industrial regions themselves and also to the east of those areas where the pollutants drifted due to the effect of the prevailing westerly winds. This phenomenon of the spread of the melanic form of *B. betularia*, due to industrial pollution, became known as 'industrial melanism'.

Kettlewell also demonstrated that the change in gene frequencies was due to natural selection and that the mechanism of selection involved differential predation by birds. This was shown by two kinds of experiment.

In mark–release–capture experiments, laboratory-reared moths were marked with spots of paint on the undersides of their wings and known numbers of the two forms were released in two contrasting woodland environments. One site was near to Birmingham city centre and was heavily polluted by soot (in the 1950s), the other was in rural Dorset and was free from pollution. When samples of the marked moths were recaptured a few days later the survival of the two forms could be compared. It was found that the dark 'carbonaria' form survived about twice as well as the typical form in Birmingham, but in Dorset the 'typicals' were at a much greater advantage.

A second kind of experiment, using direct observation and cine-photography, confirmed that predatory birds were capturing moths that had been stuck onto trees. In the Birmingham wood, the 'carbonaria' moths were better camouflaged on the darkened tree bark and were taken only about half as frequently as the 'typicals'. In Dorset, the situation was the reverse, with 'typicals' merging well into the background of the lichen-covered bark and the 'carbonaria' moths being much more exposed.

These field studies provided dramatic and convincing evidence of the 'force' of natural selection in action, determining gene frequencies in natural populations. They gave an answer about what the mechanism was and made much of their impact from the demonstration that **micro-evolution** could occur quickly enough to be observed within the lifetime of a person, and does not necessarily require the centuries of time that Darwin thought were required. Kettlewell has referred to this work on the peppered moth as *'Darwin's missing evidence'*.

During the past 40–50 years, controls limiting the amount of atmospheric pollution by smoke have brought about a reversal of the environmental damage caused over the previous century, and the 'typical' form of *B. betularia* is now in the ascendant again and slowly increasing in frequency in Manchester and other industrial areas.

Sickle-cell anaemia in humans

Sickle-cell anaemia is a genetically transmitted disease of the blood caused by an abnormal form of adult haemoglobin. We have already dealt with the structure of haemoglobin and the molecular basis of the defect that causes the disease (Chapter 12, see page 140). The gene that controls this character is inherited as a single Mendelian recessive, with two alleles and three genotypic classes. Designation of the gene is **Hb** (for haemoglobin) and the alleles are Hb^A (normal haemoglobin A) and Hb^S (haemoglobin S). The three genotypes and their corresponding phenotypes are:

1 $Hb^A Hb^A$ – normal homozygote
2 $Hb^A Hb^S$ – heterozygous carrier with normal phenotype (except under conditions of oxygen deficiency)
3 $Hb^S Hb^S$ – recessive homozygote with sickled cells.

Red blood cells of the afflicted homozygotes (Hb^SHb^S) have a characteristic crescent or 'sickle' shape (Figure 16.4). They become sticky on their surface and clump together to cause interference with the blood circulation – they are rapidly destroyed, leading to anaemia, damage to the vital organs and eventually death. Four out of five sufferers die in childhood and many of the remainder soon afterwards. In the fetus, and in very young children, there is another form of haemoglobin (fetal haemoglobin) that is not affected by the same mutation.

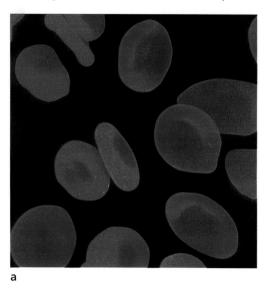
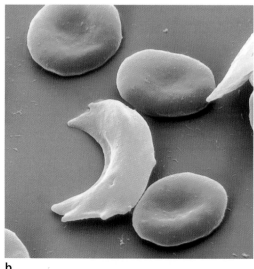

a   b

Figure 16.4 Photomicrographs of red blood cells taken on a scanning electron microscope: **(a)** normal red cells at a magnification of ×5000; **(b)** Sickle-cells ×5000

The surprising feature about this deleterious gene is that it occurs in very high frequencies (10–20%) in certain parts of the world – notably tropical West Africa, some Mediterranean countries and parts of India – and the question is, why?

In the 1950s, Tony Allison found the answer. He realised that sickle-cell anaemia is only prevalent in areas of the world where malaria is **endemic** (regularly found), and that there is a link between the two diseases. What happens is as follows.

1 The Hb^SHb^S homozygotes are at a disadvantage due to their high mortality from sickle-cell anaemia, and from a variety of other diseases to which they are also prone.
2 Homozygotes with normal haemoglobin (Hb^AHb^A) are used as a source of food by the females of various species (about 30) of blood-sucking mosquitoes (*Anopheles* spp.), which transmit the unicellular protozoan parasite *Plasmodium falciparum*. The parasite is the real predator; it undergoes part of its life cycle within the red blood cells where it multiplies, bursts the cells, and gives rise to malaria fever.
3 Heterozygotes (Hb^AHb^S) have a marked advantage over both classes of homozygotes because they are normally free of sickle-cell anaemia and they also have resistance to the malarial parasite. They have a mixture of both kinds of

haemoglobin with sufficient normal molecules to enable the red cells to function properly, but their cells are an unsatisfactory source of nutrition for the parasite. This heterozygous advantage leads to what is known as a **balanced polymorphism**. The heterozygote has the highest fitness level (that is, the highest chance of survival) and therefore maintains both forms of the gene within the population, balancing out their loss through mortality and reproductive failure of the two homozygotes. The balance is in favour of the normal form of the gene (Hb^A) because malaria has a smaller effect upon the fitness of an individual than does sickle-cell anaemia.

In this example, we can see how natural selection can act to maintain certain gene frequencies, and to preserve variation by keeping a deleterious gene within a population. If malaria is eradicated from a region, or if some members of the population move out of an area where it is endemic, then the heterozygotes lose their advantage. The polymorphism will then become *transient* and selection will gradually change the gene frequencies in favour of the normal allele.

Heavy metal tolerance in grasses: selection for a quantitative character

In discussing natural selection so far we have confined ourselves to examples involving major gene variation. Selection also acts on quantitative characters, but this is much more difficult to study. The way in which we deal with it is described in Box 16.3.

We will consider just one example of natural selection acting upon continuous variation – that of heavy metal tolerance in grasses.

Waste materials forming the spoil heaps of old derelict mine workings, where ores were once extracted, contain high concentrations of certain heavy metals such as lead, zinc, copper and nickel, which are toxic to plants and animals. These old mines are common throughout Britain and the spoil heaps are often bare and devoid of vegetation even though some of them have been in existence for 100 years or more. The mines are a good hunting ground for ecological geneticists because they represent an extreme form of environment and on some of them a few species, particularly grasses of the genera *Agrostis* and *Festuca*, have evolved metal-tolerant races that are able to colonise these barren and contaminated sites. Breeding experiments have shown that the tolerance is genetically determined and is inherited as a quantitative character, showing many grades of tolerance. Grasses may become tolerant to several metals at the same time, when these are present together; but when this happens the tolerances to the individual metals are highly specific and genetically independent of one another.

In relation to natural selection one of the most interesting features of these metal-tolerant populations is how small an area they cover, and how they manage to maintain their adaptation in the face of a continual 'inflow of genes' from the non-tolerant populations that completely surround them.

The situation is exemplified very well by the work of McNeilly on the copper tolerance of *Agrostis tenuis* growing on a small mine in Caernarvonshire in North Wales. The situation is described in Figure 16.5, overleaf. To study the adaptation of

211

the grasses, McNeilly has made comparisons between the copper tolerance of the adult plants growing on the mine area with that of their progenies raised from seed. The findings can be summarised as follows.

1 Adult plants sampled on the mine, after selection has taken place, show a high mean index of tolerance and a small range of variation. Their seeds, grown in normal soil and therefore not subject to selection, show a mean index of tolerance that is much reduced, and a wider range of variation. This indicates that:

   • the genetic composition of the seed generation has been influenced by migration of genes (carried in pollen) into the mine area from upwind non-tolerant populations
   • on the mine itself there must be some strong directional selection imposed upon the seedlings at each generation, which eliminates the low-tolerance phenotypes and maintains the adaptive characteristics of the adult population.

2 Seeds produced by copper-tolerant plants that have been grown in isolation, under experimental greenhouse conditions, maintain a high mean level of tolerance but are much more variable than adult plants growing on the mine. Because the plants that produced these seeds were grown in isolation they were protected from the pollen of non-tolerant plants, and from an inflow of their genes. Their variation is greater because they are unselected and they display variability, which results from the segregation and recombination of their polygenes.

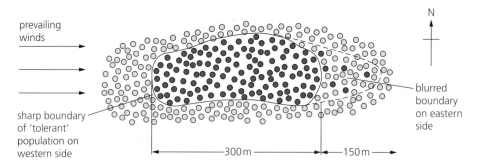

Figure 16.5 Simplified diagram of McNeilly's copper mine in Caernarvonshire showing the distribution of tolerant (dark green circles) and non-tolerant (light green circles) plants. The continuous line represents the mine boundary. At the upwind end of the mine, the boundary between the tolerant and non-tolerant populations is very sharp, while at the downwind edge, it is blurred and tolerant plants extend in to the normal population for up to 150 m

Strong selection pressure for metal tolerance, at the seedling stage, is therefore responsible for maintaining the adaptation of the mine populations and their sharp demarcation from adjacent non-tolerant plants. The overspill of tolerance genes into the surrounding downwind area occurs because selection against tolerant plants growing in normal soil is very much weaker. To some extent the gene flow from non-tolerant plants is also offset by reproductive strategies – plants in the mine populations flower earlier than those outside, and there is also a certain amount of self-pollination.

Box 16.3 Selection for quantitative characters

Genetic variation controlling quantitative characters is just as much affected by natural selection as that due to major gene polymorphisms, but the process cannot be described in the same way in terms of changes in gene frequency. With continuous variation the relation between the gene and the character is much less obvious: individual genes cannot be identified and the population geneticist has to monitor progress of selection using means and variances (variation about the mean) of the population. Three main modes of selection for polygenes have been identified and described by Mather: they are illustrated in the graphs in Figure 16.6.

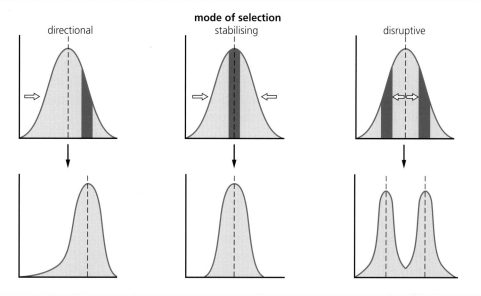

Figure 16.6 The three models of selection acting upon continuous phenotypic variation. Horizontal arrows represent the direction in which selection is operating and the dark areas show the new optimum phenotypes that will contribute to the next generation. Dashed lines represent the old and new means

Directional selection operates in a changing environment and results in a reduction in variance and progressive shift in the population mean for the character concerned until a state of adaptation is reached. It is the main form of selection practised by humans in the improvement of domesticated plants and animals.

Stabilising selection occurs when the strongest selection pressure is against the extreme variants furthest from the mean; it leads to a reduction in variance without any change in mean. It is the form of selection that operates in a constant environment to maintain the best adapted genotypes within the population.

Disruptive selection favours two optimum classes at the expense of intermediates. It occurs in natural populations where two different habitats, or kinds of resources, exist, and in plant and animal breeding where selection is practised for extremes of size and form.

Other forces of evolutionary change

Gene mutations are rare events, occurring at the rate of about once in every 100 000 gametes for a given locus (Chapter 13). Mutational change can alter allele frequencies in populations but the degree of change is thought to be quite negligible in comparison with other forces such as selection.

Migration (gene flow) is the transfer of genes between populations by the movement of gametes, individuals or groups of individuals from one population to another. Its most important effect is to counteract local adaptation and divergence, by introducing genes from one population into another one of different genetic make-up.

Genetic drift is the random fluctuation in gene frequencies due to sampling error. In a large population, there is a strong likelihood that the sample of genes coming together in the zygotes will be fully representative of the gene pool in the previous generation. A small population on the other hand is much more prone to sampling error, and the genes taken into the zygotes may be quite *unrepresentative* of the gene pool, due to chance alone. In a small population, therefore, the gene frequencies may fluctuate widely from one generation to the next.

The **founder effect** is a once-and-for-all change in gene frequencies that may take place when a small group of individuals wanders off, or becomes isolated, from the main population and 'founds' a new population. Because the group is small it may carry an unrepresentative sample of genes, due to sampling error, and the new population may differ widely in its genetic constitution from the original one. This can happen in the formation of island races, for example, or when the numbers of a population are suddenly reduced by some catastrophe. It is the most drastic way that is known of bringing about a sudden change in gene frequencies. Descendants of races formed in this way can differ from their ancestral population in many gene loci.

Speciation

There are almost as many definitions of the term 'species' as there are different species. Problems of definition arise because in taxonomic terms a species means organisms that share specific morphological characters, and they are often described from a few dead 'type specimens'. In nature, however, species vary in space and time, and there are often gradations in form leading to uncertainty about where one species ends and another one begins. The song sparrow (*Melospiza melodia*), for instance, is widely distributed throughout the United States and there are many local **races** that each have their own distinctive form and song pattern. Are they all members of the same species? The answer is yes, because, just as in humans, wherever individuals from different geographical races come together they will interbreed and give rise to intermediate populations. It is this capacity for interbreeding that unites organisms into species and which separates one species from another. Races will interbreed but species will not.

We can therefore define a species as a group of interbreeding natural populations that share common morphological characteristics and are reproductively isolated from other such groups. This definition obviously cannot apply to self-fertilising organisms, or to forms that reproduce solely by asexual methods. Nonetheless, such groups are frequently called species where they consist of individuals that are very similar to one another. The species is the largest group of organisms that share a common gene pool. There are about three million such groups on Earth at the present time, but their number and kinds are slowly changing all the time. Some of these groups are in a stable relationship with their environment, some are becoming extinct, and others are in the process of evolving into new reproductive groups – that is, undergoing **speciation**.

Mechanisms of speciation

It is not possible to give a clear-cut and definite account of the process by which new species arise. In some special cases, such as the 'instant speciation' that may result from polyploidy in plants (see page 219), the mechanism is well understood. But the majority of species do not evolve in such a simple or rapid way. The process is usually gradual and takes place over a long period of time – of the order of several thousands of years.

The general idea about speciation is that natural selection, and the other forces of evolutionary change mentioned above, may act differently in the different populations that comprise a species. When such differential action continues over a large number of generations the populations concerned may slowly diverge from one another, by the accumulation of numerous small genetic differences, and begin to form a number of different *races*. This process occurs because no two populations can occupy precisely the same niche, or exploit exactly the same kind of resources at the same time. As long as there is interbreeding between the populations, however, they will continue to exchange their genes and will remain united as one species. The critical event that allows the divergence to proceed beyond the level of races, and into new species, is the formation of some kind of *barrier* that prevents gene flow between populations. Such a barrier preventing gene exchange is called an 'isolating mechanism'. There are a number of different isolating mechanisms, which are listed in Figure 16.7 (overleaf), and therefore several different ways in which divergence and speciation may occur.

In discussing the mechanism of speciation, it is customary to distinguish between that which occurs due to some method of physical separation of the gene pool into different geographic regions (**allopatric speciation**) and that which arises from an isolating mechanism within a gene pool in the same geographic region (**sympatric speciation**).

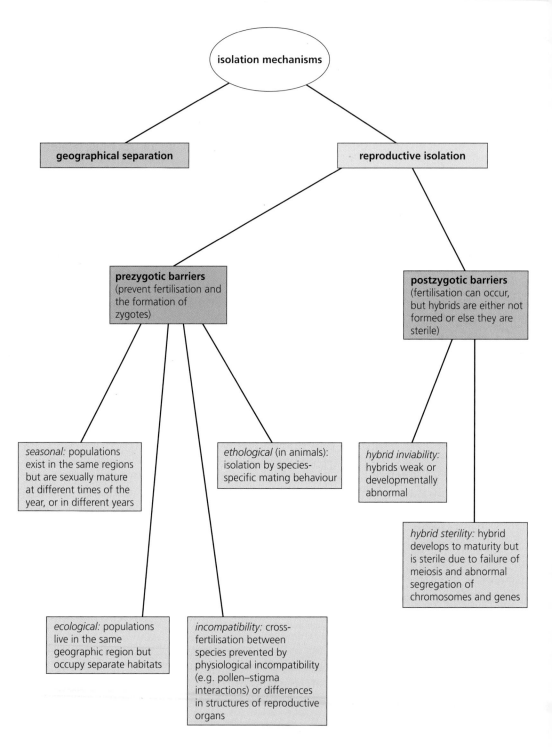

Figure 16.7 A summary of the most important mechanisms that isolate closely related organisms into separate breeding groups

We will confine ourselves here to discussing a relatively simple model of speciation – geographic speciation (Figure 16.8).

1 The initial event involves some physical separation of an existing population into two or more sub-populations so that the free exchange of genes is prevented. This may occur, for example, when part of a land mass becomes detached as an island.

2 Once spatially separated, the sub-populations will diverge genetically. Their environments will be different, they will have different patterns of variation, and selection will act in different ways to alter their genetic composition.

3 Eventually this divergence may proceed to such an extent that the separate groups are no longer able to interbreed even if they are brought close together again as sympatric populations. At this stage they are reproductively isolated and by definition have become different species.

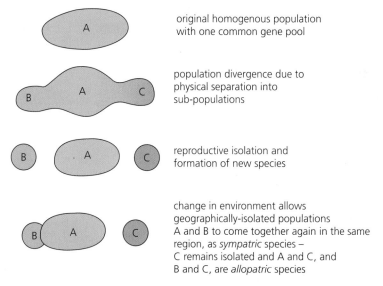

original homogenous population with one common gene pool

population divergence due to physical separation into sub-populations

reproductive isolation and formation of new species

change in environment allows geographically-isolated populations A and B to come together again in the same region, as *sympatric* species – C remains isolated and A and C, and B and C, are *allopatric* species

Figure 16.8 Diagram showing the sequence of events that leads to geographic speciation, starting with a homogenous group of organisms (A) sharing a common gene pool

Darwin's Finches: an example of geographic speciation

The Galapagos Islands were visited by Darwin in 1835 during the voyage of H.M.S. *Beagle*. They are a group of volcanic islands that were pushed up out of the sea more than a million years ago, and have never been connected to the mainland of South America. They lie 600 miles west of Ecuador. The land animals that inhabit them must be descendants of individuals that came across the sea from the mainland. Among the few species that have established themselves, there are a number of finches. Specimens of these birds were collected by Darwin and have subsequently been studied in great detail – notably by David Lack who spent a year on the Galapagos (1938–39). In all there are 14 species, 13 on the Galapagos and one on nearby Cocos Island: collectively, the birds are known as Darwin's Finches.

Darwin was greatly influenced by the variety in these finches, particularly by the different sizes and shapes of their beaks that are associated with their different habits of feeding (Figure 16.9). He realised that the islands must have been colonised by migrants from the mainland, where there is only one type of finch, and that their descendants had become diversified and adapted to the various ecological niches that they found. **Adaptive radiation** is the term now used to describe the evolution of a group of organisms along several different lines involving adaptation to a variety of environments.

Figure 16.9 Some of Darwin's Finches, which originated on the Galapagos Islands from a common ancestor through adaptive radiation. They differ mainly in the form of their beaks and have different feeding habits that are associated with the different ecological niches available on the Islands.
(a) Insect-eating tree finch. **(b)** Vegetarian tree finch with parrot-like beak for eating fruits and buds.
(c) Warbler-like finch feeding on small insects. **(d)** Cactus-eating ground finch (the other ground finches are mainly seed eaters). **(e)** Woodpecker-like finch, which climbs trees and uses a cactus spine to dislodge insects from crevices in the bark. When the insects emerge it drops the spine and seizes them with its beak (after David Lack)

It is envisaged that the different species evolved in isolation on separate islands, and some of them still remain isolated. Some of the others, however, now live as sympatric species on the same island. The different species co-exist in this way because they are reproductively isolated by mating behaviour, and there is no competition between them for food – they have evolved differences in diet. It is important to appreciate that this geographic speciation could take place on the Galapagos (and in other similar situations) because the ancestors of the finches were able to colonise habitats that were largely unoccupied when they arrived. On the mainland this diversity of finches is not found – here the available niches were already filled by species of other birds. The woodpecker-like finch, for example, could never have evolved on the mainland in competition with the true woodpecker.

Polyploidy and speciation

Polyploidy is an important exception to geographic speciation, and is a means of bringing about reproductive separation in sympatric populations. It is much more important in plants than in animals, as discussed in Chapter 14.

When chromosome doubling takes place in a diploid, to give an autotetraploid, the newly formed polyploid will immediately be isolated from its parent species by a sterility barrier. The reason for this is that hybrids between the new 4x and the original 2x forms will be triploid (3x), and these are sterile due to irregular chromosome behaviour at meiosis (Chapter 14).

In allopolyploidy, a new species can arise by hybridisation between two existing species, followed by spontaneous chromosome doubling (Chapter 14): again it will be reproductively isolated from both diploid ancestors due to triploid sterility. Cultivated wheat is an important example of the origin of a new species by allopolyploidy (Box 14.1, page 179). Another well-known case is that of cord grass, *Spartina townsendii*. This species is abundant around the coasts of Britain where it has been widely used to stabilise mud flats. It is thought to have arisen in about 1870 in Southampton Water, by hybridisation between an introduced American species (*S. alterniflora*) and a native British species (*S. maritima*) to form a sterile hybrid *S. anglica*. Doubling in chromosome number gave rise to the fertile allotetraploid *S. townsendii*.

Summary

◆ Species are not immutable. They undergo continuous gradual change and evolve into new forms.

◆ The principal agent of change is natural selection, which acts like a sieve and favours those forms that are best adapted to their changing environments.

◆ Genetics explains how evolutionary change takes place, and strengthens the theory of evolution by natural selection.

◆ Neo-Darwinism can explain the source of variation within a species, and can show how the genetic composition of natural populations can change over short periods of time due to micro-evolution.

◆ The study of population genetics has also revealed that there are forces of change other than natural selection: the principal one is genetic drift.

◆ Speciation is the result of natural selection acting in different ways upon homogeneous populations that become separated into different breeding groups. This separation comes about, and is maintained, by various methods of reproductive isolation, which provide barriers to gene exchange.

Further reading

1 R. J. Berry (1977) *Inheritance and Natural History*. Collins.
2 A. D. Bradshaw and T. McNeilly (1981) *Evolution and Pollution*, Studies in Biology No. 130, Edward Arnold.
3 R. Dawkins (1989) *The Extended Phenotype*. Oxford University Press.
4 R. Dawkins (1996) *River out of Eden*. Phoenix.
5 R. Dawkins (1991) *The Blind Watchmaker*. Penguin Books.
6 J. Diamond (1992) *The third chimpanzee: the evolution and future of the human animal*. HarperCollins, New York.
7 E. B. Ford (1973) *Evolution Studied by Observation and Experiment*, Oxford Biology Reader Series. Oxford University Press.
8 D. J. Futuyama (1998) *Evolutionary Biology*. Third edition. Sinauer.
9 C. Patterson (1998) *Evolution*. Second edition. Natural History Museum.
10 P. W. Price (1996) *Biological Evolution*. Saunders.
11 P. M. Sheppard (1975) *Natural Selection and Heredity*. Hutchinson.
12 G. L. Stebbins (1982) *Darwin to DNA, Molecules to Humanity*. Freeman.
13 B. Wood (1996) 'Human Evolution', *BioEssays* 18: 945–954.

Questions

1 Describe the theory of evolution put forward by Darwin and some of the evidence he used to formulate his theory. How has this theory been modified in the light of modern biology?

2 Critically evaluate the evidence that natural selection has been the main force of evolution.

3 **a** What is a species?
  b Why is isolation important for the evolution of new species? By what mechanisms can species become isolated?

4 What is meant by the term *genetic polymorphism*? Give one example of a transient polymorphism and one example of a stable polymorphism. What are the main factors that influence the polymorphisms in your examples?

5 Snails of the genus *Discula* are found on one of the Canary Islands. They are thought to have originated from a founder species brought on to the island either by birds or on logs floating on the sea. The genus has since split into two subgenera, each with two species. Species **A** and **B** belong to subgenus one, **C** and **D** to subgenus two.
  a What is the meaning of the terms *genus* and *species*?
  b Species **C** and **D** inhabit different mountains on the island. Suggest why the species show different gene frequencies for many of the genes they have in common. Explain how these differences have arisen.
  c Why might competition between species **A** and **B** be greater than that between species **A** and **D**?

6 Read the following passage.
  Madagascar
  The island of Madagascar has been described as the laboratory of evolution. It broke away from mainland Africa at least 120 million years ago and, following this, many new species developed. Estimates of the number of plant species on the island vary from 7370 to 12 000, making it botanically one of the richest areas in the world. Of 400 flowering plant families found world-wide, 200 grow only in Madagascar. Among animals, true lemurs are found nowhere else, and 95% of the country's 235 known species of reptiles evolved on the island. Over the past 25 years the human population has doubled. Land shortage is leading to clearing of the forest, and Madagascar is now facing deforestation on a massive scale. Scientists have estimated that even if the forest could recover, regeneration could take up to a hundred years.

  (Adapted from an article in New Scientist)

  a Explain what is meant by the term *species*. [2]
  b **(i)** Explain the processes which might have led to the evolution of new species on the island. [4]
  (ii) Suggest and explain how the number of animal species on the island may have changed as it became '*botanically one of the richest areas in the world*'. [2]
  c Describe the processes by which forest is able to regenerate after being cleared. [4]
  AQA (NEAB) AS/A Biology: Continuity of Life (BY02) Module Test March 1999 Q8

7 Fourteen different species of finch live on the isolated Galapagos Islands. The finches are believed to have all evolved from a single common ancestor. The diagram shows the suggested evolutionary relationships among the species.

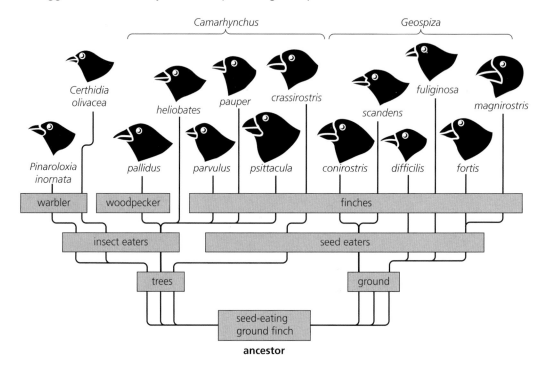

a Suggest why these fourteen types of finch are considered to be different species. [1]

b Use the information in the diagram to identify:

(i) the species of ground finch which eats the largest seeds

(ii) the number of genera which are present in the islands. [2]

c Suggest how two distinct species, one insect-eating and the other seed-eating, may have evolved from a common ancestor. [4]

AQA (NEAB) A Biology Paper 1 June 1999 Section B Q14

Selection imposed by humans

In recent times, human activity has had a profound effect upon the population genetics of other species. Large areas of the Earth's land surface are now covered with domesticated forms of plants and animals. Wild populations are frequently subjected to various forms of chemical treatment aimed at their control, or else at their destruction and eradication. This applies particularly to species such as disease-causing organisms, which are directly harmful to humans, and to those that are competing indirectly with us for sources of nutrition: for example, fungal pathogens of crop plants; weeds competing with crops for water, light and nutrients; animals that attack stored food products; predators of domestic animals.

In this chapter, we consider some aspects of selection imposed by humans. The distinction between what is said here and some of the examples of selection given in Chapter 16 is somewhat arbitrary. Obviously the selection that gave rise to the spread of melanic moths was the result of human activities, although not deliberate. Changes in the genetic composition of species and populations included in this chapter are the result of deliberate interference with other organisms for our own gain and protection – although these changes are not necessarily all harmful. It is perfectly possible to domesticate and to breed from a wild species of plant or animal without unduly disturbing the natural populations from which it came.

Selection imposed by humans is often more rapid and intense than that which occurs in nature. The selection pressures are very strong and can be applied to produce extremes of diversity and development. The results of these activities demonstrate the wealth of genetic variation that is available within a species and how it can be manipulated to produce large changes in the form and physiology of organisms. The variation that we see under domestication provides strong support for the theory of evolution by natural selection.

Domesticated plants and animals

All the forms of plants and animals that we use today for our food, and various other resources, are derived from wild species. The vast majority were domesticated several thousand years ago and since that time they have undergone a phase of rapid evolutionary change. Until the twentieth century, the selection was imposed 'unconsciously', in that our ancestors had no idea of what they were doing in terms of genetics. The selection, and 'breeding', of plants and animals was an art, rather than a science as we know it today. Scientific principles were only applied after the laws of heredity had been established, and when an understanding was gained of how to handle characters that show continuous variation.

Plants

The early phase of 'unconscious' selection in plants was so successful that there have been almost no new staple crops introduced in the world for the past 2000 years. Most of the major cereals, root crops, fruits and vegetables were established before that time, although their improvement has been going on continuously to the present day. An important exception is *Triticale* – a new allopolyploid cereal produced in the twentieth century by hybridisation between wheat and rye.

The early phase of domestication was little more than natural selection – saving seeds from the biggest and best individuals, and planting out the ones that survived the ravages of various pests and diseases. Selection was done on phenotypes, of course, but over the centuries many small heritable variations accumulated and gave rise to some remarkable changes. The major part of the selection was on characters that show continuous variation, but some progress was also due to major mutational events like the allopolyploidy that gave rise to our cultivated wheats (Box 14.1, page 179). One example will serve to illustrate the range, and kind, of variation that can be obtained by selection from within a species: that of different cultivated forms derived from the wild cabbage (Figure 17.1).

Figure 17.1 Variation under domestication in the cabbage, *Brassica oleracea*. The wild cabbage is shown top left, **(a)**. The other photographs show cultivated forms of the species, which have been modified by selection for different parts of the plant. The cauliflower **(b)** is a modified inflorescence comprising thousands of fleshy flower buds. Brussels sprouts **(c)** have been selected for axillary buds and the cabbage **(d)** for the numerous overlapping leaves which surround the terminal bud. Three other varieties are also shown – broccoli **(e)**, kale **(f)** and kohlrabi **(g)**. If they are left to flower, all the different forms can be hybridised

Animals

The first animal species to be taken into domestication by man was the dog (*Canis familiaris*). Remains dating from 10 000 BC have been found in Israel and Iraq. The dog is descended from the wolf and became our most widely distributed domestic animal. It was found to be useful for hunting, keeping flocks of animals, scavenging and also as a source of food. At the present time its main role is as a popular pet.

There are now more than 500 distinct breeds of dog, which have evolved over a period of time by selective breeding to suit different requirements. They all belong to the same species and can be interbred to give mongrels. The dog is probably the best example to show the extent of variation that can be produced by selection within a species of domestic animal – we have only to compare the St Bernard and the Chihuahua to appreciate the point. The way in which the anatomy has been modified by selection for different breed characters is illustrated in Figure 17.2. Changes in the skull are particularly striking – no other species of animal shows such a large divergence in this character.

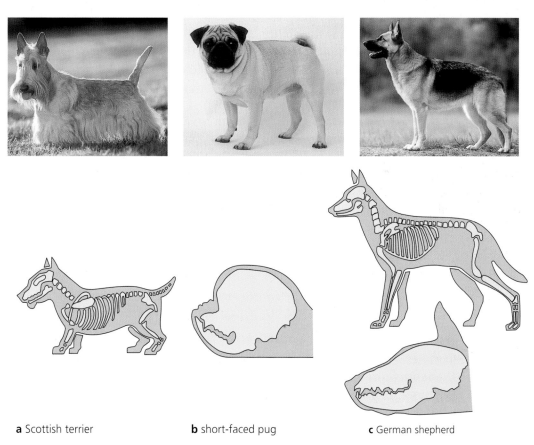

a Scottish terrier **b** short-faced pug **c** German shepherd

Figure 17.2 Effect of selection on skeletal structure in the dog. Note the change in the leg bones following selection for miniaturising features in the Scottish terrier **(a)**; and the alteration in the nasal bones of the short-faced pug **(b)** compared to the German shepherd **(c)**

Resistance to pesticides and antibiotics

Humans are engaged in a constant struggle against other species that compete with us for food and other natural resources. An enormous part of the potential yield of our crops and animal products is lost each year to pests of one sort or another (various animals, insects, fungi, bacteria, weeds). To minimise these losses we combat the species concerned with a wide range of control measures. One of the most common lines of attack is the use of chemical poisons, called 'pesticides'. The term 'pesticide' is very general and includes any substance used to control organisms that cause injury, disease or losses in growing crops, domestic animals or humans. We can also refer to particular kinds of pesticides, such as insecticides, herbicides, fungicides, nematocides, rodenticides, bactericides, and so on.

An **antibiotic** is a substance, produced by a micro-organism, that is able to inhibit the growth of other microbes, or to destroy them. Antibiotics may be purified from the organisms that produce them naturally, or made by chemical synthesis. Many of these chemical controls are only partly effective, however, because in time it is found that their target organisms respond by evolving resistant strains that can overcome them.

Resistance, in this context, is the ability of organisms to survive exposure to a normally lethal dose of a chemical poison. Resistance develops because a certain proportion of individuals already possess resistance alleles to many different chemicals, even though they have never been exposed to them. This is a result of random changes brought about in genes by mutation. The chemicals themselves do not cause mutations; they select out the resistant individuals, which then survive to give progeny. The genetic composition of the pest population, and many other 'innocent' species, thus changes in response to selection imposed by humans. There are today over 500 known cases of insect species that have developed resistance to insecticides, and somewhat lower figures for fungicides, bactericides and herbicides. The resistant forms have all become apparent since the 1950s, as a result of selection pressures imposed by the frequent application of high levels of pesticides. Some examples are detailed below.

Warfarin resistance in rats

Warfarin (3-α[acetonylbenzyl]-4-hydroxycoumarin) is an anticoagulant poison widely used as a 'rodenticide' for the control of rats and mice. It was first introduced in 1950 and became a popular poison, against rats in particular, because of its low toxicity to farm animals. It acts by interfering with the way in which vitamin K is utilised in the normal process of blood clotting. When the bait is administered to sensitive rats the capillaries become much more fragile than normal, and the blood fails to clot. The affected animals succumb to haemorrhages and slowly bleed to death – apparently without pain!

Resistant strains of rat first appeared in Scotland in 1958. By 1972 resistant rats had appeared in twelve other areas of Britain. Breeding tests established that its genetic basis could be explained in terms of a single gene mutation in which

226

resistance is conferred by a dominant allele. The gene is designated as **Rw** and its dominant and recessive alleles as $\mathbf{Rw^r}$ and $\mathbf{Rw^s}$ respectively. Evidently, the blood clotting process of the resistant strains is insensitive to warfarin: clotting takes place in the normal time, but there is an enhanced requirement for vitamin K.

One resistant population that has been particularly well monitored is that in an area centred on Welshpool, in mid Wales, and extending over the border into the English county of Shropshire (Figure 17.3). Resistant rats were first noticed near Welshpool in 1959. Thereafter they were seen to spread out from the original site at a rate of three miles per year, which is the normal rate at which rats invade new territory. Their progress depended on the continued use of the poison, and the constant selection pressure that was applied to the population, for several years. What happened here was that the environment of the rat, in terms of diet, was suddenly and deliberately changed by the intervention of human activities. This altered the selective value of the genes normally involved in the blood clotting process, and their action suddenly became lethal in the homozygous form ($\mathbf{Rw^s Rw^s}$). A few variants carrying the dominant resistance allele were then favoured by selection, in the new environment, and the genetic composition of the population changed in response to the warfarin baiting.

In any wild population of rats mutation constantly generates new alleles of the genes that mediate in the biochemistry of blood clotting – and some of them will confer resistance to warfarin even though the population has never been exposed to the substance previously. Under normal circumstances these mutations would most likely be deleterious, as indeed they are in the case of rats in a natural environment, and selection would keep them down to mutation frequency.

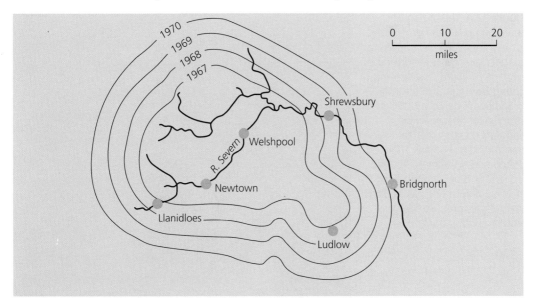

Figure 17.3 Map of the area around Welshpool where warfarin resistance in rats has been intensively studied. The map shows the spread of resistance over the three-year period 1967–70 (based on J. H. Greaves, and B. D. Rennison, 1973, *Mammal. Rev.* **3**, 27–29)

The genetics of warfarin resistance turns out to be more complicated, and more interesting, than it appeared at first sight. The complication arises from the fact that rats that are homozygous for the dominant resistance allele have a requirement for massive amounts of vitamin K, twenty times more than normal, which they have difficulty in meeting (vitamin K is synthesised by the gut bacteria). Normal homozygotes are poisoned by warfarin, while the heterozygotes survive and require only a fraction more vitamin K than normal. Genotypes and phenotypes are as follows.

1 Rw^sRw^s – normal rats, sensitive to warfarin (recessive phenotype).
2 Rw^sRw^r – heterozygotes, resistant to warfarin and having slightly increased requirement for vitamin K.
3 Rw^rRw^r – resistant to warfarin but having a twenty-fold increase in their vitamin K requirement, which greatly reduces their fitness compared to the heterozygotes.

The heterozygotes therefore have a selective advantage over both classes of homozygotes in the areas where warfarin is used, and this results in a balanced polymorphism – as in sickle-cell anaemia (Chapter 16). Susceptibility to warfarin in the recessive homozygotes (Rw^sRw^s) is balanced by the reduced fitness of the resistant homozygotes (Rw^rRw^r). Because of this the normal recessive allele can never be completely removed from the population.

Following the rapid spread of warfarin resistance in the 1970s, a new generation of related anticoagulant poisons was introduced. Difenacoum is one of the most widely used at the present time, but resistant populations are again emerging in some areas of Britain. Optimistic press reports declaring 'High Noon for Rats' are being superseded by more realistic headlines proclaiming 'Super Rats Overcome New Poison'.

Figure 17.4 The brown or common rat (*Ratus norvegicus*) is the most widely distributed of all rodents. It causes enormous damage by eating and spoiling both growing and stored food. In addition it damages the structure of buildings and harbours diseases that are transmitted to humans and their domestic animals. The rat is a prolific breeder and under ideal conditions one pair could theoretically produce 3000 individuals in one year

DDT resistance in insects

This is the best known case of resistance to an 'insecticide'. DDT (dichlorodiphenyltrichloroethane) was introduced in the 1930s. It is cheap, easy to make and a relatively stable substance (Figure 17.5).

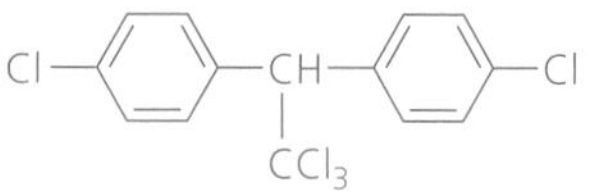

Cl—⟨⟩—CH—⟨⟩—Cl
CCl_3

Figure 17.5 Structural formula of DDT

On a non-porous surface, DDT will remain active for as long as eighteen months. In insects its main mode of action is on the nervous system: it causes loss of movement followed by paralysis and death. There are several mechanisms of resistance, determined by a number of different genes:

1 the nervous system itself may have a reduced toxic response to DDT
2 the insect cuticle may be less permeable to DDT
3 there may be an altered pattern of behaviour, so that the resistant strains avoid contact with the chemical
4 there may be an increase in the insect's lipid content enabling the fat-soluble DDT to be separated from sensitive parts of the organism
5 some insects carry high levels of an enzyme that detoxifies DDT.

In the housefly (*Musca domestica*), it is thought that many resistant strains contain high levels of the enzyme DDT-dehydrochlorinase, which breaks down the DDT into products that are less toxic (detoxification). Natural populations are pre-adapted in that there are always a few variants that carry high levels of the enzyme – even though they have never encountered DDT. Resistance is therefore the result of selection for these variants – they have higher levels of fitness and leave more progeny than the susceptible genotypes. Other important pests with resistance to DDT are the yellow fever mosquito (*Aedes aegypti*), and various species of malaria-carrying mosquitoes (*Anopheles* spp.).

Enormous quantities of DDT have been used over the years and much has accumulated in the fat deposits of animals for which it was never intended. It gets into food chains and then becomes concentrated in the bodies of predatory animals, including humans. The use of DDT has now been abandoned in many parts of the world.

Insecticide resistance in peach-potato aphids

Peach-potato aphids (Figure 17.6, overleaf) have developed resistance to several different insecticides used against them in fields and greenhouses. They display what is known as **cross-resistance** to organophosphorus, carbamate and pyrethroid insecticides. Resistance that develops in response to any one of these substances will give protection against the others, and it appears to be based on greater activity of the enzyme carboxylesterase (E4). The form of the enzyme is the same in both resistant and susceptible strains, but the resistant ones produce much more of it.

Figure 17.6 The peach-potato aphid, *Myzus persicae*

E4 degrades insecticides (and other substances) by hydrolysing them. There are variants of aphids that differ in the concentration of the E4 enzyme that they carry, and these differences appear to be due to a series of *duplications* of the structural gene for the enzyme. The most resistant of the variants has 64 times as many copies of the gene as the susceptible strain, which has only one copy per haploid chromosome complement. Amplification of genes coding for E4 confers a selective advantage on the aphids in the presence of the insecticides. In Britain, susceptible strains have been almost entirely replaced by the resistant variant, which produces four times the normal quantity of E4. The most resistant variants with a 32 and 64-fold increase in E4 are mainly confined to greenhouses, where insecticides are used much more intensively.

Resistance to antibiotics in bacteria

Bacteria are important pathogenic organisms in humans, and in domesticated animals. In humans they are responsible for such serious diseases as tuberculosis, dysentery, urinary infections, typhoid fever and cholera, to mention but a few. Following the introduction of antibiotics, starting in the 1940s, it was thought that many of these diseases would be quickly eradicated. This was not the case, however, and resistant strains of these disease-causing organisms quickly developed in response to the widespread use of antibiotics.

Some of the most troublesome bacteria are *Staphylococcus* spp., which live in all our noses and are the cause of many serious infections – particularly in hospital surgical wards. Penicillin was highly effective against this bacterium when it was first brought into clinical use in 1940. Within ten years, though, the majority of staphylococcal infections in hospitals throughout the world were penicillin resistant. The basis of the resistance is the enzyme penicillinase, which cleaves the β-lactamase ring of penicillin and renders it inactive. The β-lactamase enzyme, and the gene that codes for it, have always been present in bacterial populations (penicillin is a natural substance after all). Strains that produced high levels of penicillinase were selected for when penicillin became widely used. In the 1950s new antibiotics were introduced, such as streptomycin, chloramphenicol, tetracycline, and so on, but these in turn have all led to the emergence of resistant strains.

Drug resistance is becoming a serious problem world-wide. We are running out of antibiotics that can be used for the treatment of some pathogens, which are evolving resistance faster than we can develop new antibiotics. In Spain, for example, the over-prescription of penicillin has led to perhaps 80% of bacteria being resistant to the drug. There is now a need to limit the widespread use of antibiotics for non-serious conditions. In particular, physicians are becoming more reluctant to prescribe antibiotics for minor infections, especially for colds and flu, which in any case are caused by viruses rather than by bacteria and so cannot be cured by antibiotics.

Many pathogens now have multiple drug resistance; that is, they are resistant to many antibiotics that have been used to try to kill them. Some remain susceptible only to the drugs vancomycin and teicoplanin; for example, *Mycobacterium tuberculosis*, which causes tuberculosis, and *Streptococcus pneumoniae*, which causes pneumonia. The recent emergence and increasing prevalence of vancomycin-resistant strains poses a serious clinical threat because no other effective antibiotics are now available.

Bacteria frequently carry **plasmids** as well as their main chromosome. These plasmids are small circular molecules of DNA that can be classified into several types. Plasmids of one class, known as **R-plasmids**, carry genes for drug resistance, although not all resistance genes are necessarily carried in plasmids. Plasmids are self-replicating structures that are inherited when the bacterial cells multiply by binary fission. They can also be transferred from one strain to another by conjugation, and this greatly facilitates the spread of resistance genes. **Infectious drug resistance** is the term used to describe the way in which resistance genes can be transmitted by plasmids during conjugation.

In the late 1950s, in Japan, it was discovered that some strains of *Shigella*, the bacteria that cause dysentery, carried plasmids with several different resistance genes in them; that is, they had **multiple drug resistance**. This multiple resistance had built up because the bacterial populations in hospitals had been exposed to several different drugs over a period of time. The problem is not confined to Japan, but is now common throughout the world. Because the genes concerned are in plasmids it means that they are all transferred together, during conjugation, from a multiple-resistant strain to a sensitive strain which lacks the plasmid. These R-plasmids can also be exchanged between different species of bacteria, and drug resistance can be passed, for example, from *Shigella* to the normal gut bacterium *E. coli*. One person can therefore pick up multiple drug resistance in *Shigella*, from another person, through *E. coli* as an intermediate. Multiple drug resistance built up in farm animals can also be transferred to bacteria carried by people working with them. At one time this posed a serious threat to public health, because farm animals were routinely fed on diets containing antibiotics in order to improve their growth rates. Legislation has now been passed to restrict the use of antibiotics in animal foodstuffs.

Summary

◆ Humans have deliberately interfered with populations of other species in order to feed themselves and gain protection from pests and disease-causing organisms. In the process, we have imposed selection on the species concerned.

◆ Wild plants and animals have been domesticated, over a period of several thousands of years, and changed by strong selection pressures out of all recognition from their original form.

◆ This response to human selection has provided convincing evidence of the power of selection to act on the variation within a species and to lead to diversity and evolution.

◆ The use of pesticides and antibiotics is a recent phenomenon, and its effects have mainly taken place since around 1950. In an attempt to control or eradicate pests, humans have applied massive quantities of poisonous chemicals to the environment. The target organisms have responded by evolving mechanisms of resistance. This evolution has led to changes in the genetic composition of their populations and to individuals adapted to the altered environments.

Further reading

C. B. Heiser (1981) *Seed to Civilisation.* W. H. Freeman.

Questions

1 Explain how studies on drug resistance in bacteria contribute to our understanding of evolution.

2 Write an essay about how humans have intentionally and unintentionally influenced the evolution of other species.

3 Insects often exhibit resistance to insecticides. Some houseflies produce the enzyme DDT-dehydrochlorinase, which makes them resistant to DDT. Mosquitoes quickly became resistant to another insecticide called dieldrin. Explain the mechanism by which resistance has developed. Why may the resistance of mosquitoes to insecticides have developed particularly quickly?

4 Warfarin is a pesticide which is used to kill rodents such as rats and mice. Some rodents are resistant to warfarin and such resistance was first discovered in wild rats on farms around Welshpool in 1959. Resistance to warfarin is controlled by a gene with two alleles, W^1 and W^2.

In 1959, a study was made of the genotypes of rats on farms around Welshpool where warfarin had been used. The genotypes of 74 trapped rats were determined. The results are shown in the table below. The table also shows the expected numbers of each genotype, assuming the frequencies of the two alleles, W^1 and W^2, in the population remain constant.

Genotype	Phenotype	Observed number	Expected number
W^1W^1	susceptible to warfarin	28	32
W^1W^2	resistant to warfarin	42	33
W^2W^2	resistant to warfarin	4	9

a (i) Determine whether the allele for resistance to warfarin, W^2, is dominant or recessive, giving a reason for your answer. [1]
(ii) The χ^2 test was used to determine whether the difference between observed and expected numbers is significant. A value of 5.73 was obtained for χ^2. Using the extract from a table of values (below), state whether the difference is significant, giving a reason for your answer. [2]

Probability levels $P/\%$	99	10	5	2	1	0.1
χ^2		0.00	2.71	3.84	5.41	6.64	10.83

b Rats with the genotype W^2W^2 require much more vitamin K in their diet than those with the other genotypes.
(i) Calculate the observed number of rats of each homozygous genotype (W^1W^1 and W^2W^2) as a percentage of the expected number of that genotype. Show your working. [3]

(ii) Both homozygous genotypes are at a disadvantage compared with the heterozygous genotype. Using your calculated results from **b (i)**, state which of the homozygous genotypes is at the greater disadvantage. [1]

(iii) Give a reason why each of these genotypes (W^1W^1 and W^2W^2) is at a disadvantage compared with the heterozygous genotype. [2]

London A Biology Synoptic Paper Module Test B6 January 1999 Q4

5 Farmers have steadily improved the yields from plants and animals by the process of selective breeding. One of the factors determining whether selective breeding will be worthwhile is heritability. Heritability can be defined as the proportion of all phenotypic variation in a population that is due to genetic effects.

Heritability values can range from zero (with no influence of genes on phenotypic variation) to 1.0 (where all the phenotypic variation is due to genetic differences). The table shows some estimated heritability values.

Feature	Organism	Heritability value
plant height	corn	0.70
yield	corn	0.25
body mass	chicken	0.20
egg mass	chicken	0.60
egg production	chicken	0.30

a Use the information in the table to evaluate whether it would be worthwhile to select for an increase in egg mass and body mass in chickens. [2]

b Inbreeding, the crossing together of related individuals, has been used extensively in selective breeding programmes. One effect of inbreeding is to produce strains which are homozygous at practically all gene loci. The diagram shows the effect of inbreeding at one gene locus.

	percentage heterozygosity	genotypes
parents	100	Aa
F$_1$	50	AA 2Aa aa
F$_2$	25	4AA 2AA 4Aa 2aa 4aa

(i) Explain what is meant by the term *gene locus*. [1]

(ii) Calculate the percentage of F$_3$ plants which you would expect to be homozygous if inbreeding was continued. [1]

(iii) Explain why undesirable features may appear more frequently in some strains after inbreeding. [1]

AQA (NEAB) AS/A Biology: Continuity of Life (BY02) Module Test March 1999 Q6

18

Genetic engineering

Work on the improvement of domesticated species goes on all the time, and has now reached a high level of technological sophistication. Such manipulation relies mostly on natural processes – recombination, selection, hybridisation, and chromosome doubling (for example, with colchicine). In the 1970s, however, a new development occurred that completely revolutionised the way in which genetic material could be handled: 'restriction enzymes' were discovered.

Restriction enzymes were isolated from bacteria and found to have the property of cutting DNA into discrete fragments. When they were used in conjunction with other enzymes that join up broken ends, it became possible to combine together pieces of DNA from totally unrelated species, and to make what is called 'recombinant DNA'. Recombinant DNA technology allows us to manipulate DNA in ways that are quite novel, and in some cases unnatural – such as placing human genes into bacterial chromosomes so that they function and make their gene products within the bacteria cells. This new facility has enormous potential for genetic research, as well as for the modification of organisms by 'genetic engineering'.

Genetic engineering

Genetic engineering is the production of recombinant DNA and its incorporation into a host organism, which is then known as a *transgenic* organism. Some aspects of this new branch of genetics are discussed below.

Restriction enzymes

Virtually all species of bacteria synthesise enzymes that make double-stranded cuts in DNA. They apparently use these enzymes to protect themselves from foreign DNA that gets into their cells – mainly from bacterial viruses. Because these enzymes restrict the range of host bacteria that a virus can invade, they are known as **restriction enzymes**, or more precisely **restriction endonucleases**. Restriction enzymes can be extracted and purified from cultures of bacteria and then used by the genetic engineer to make recombinant DNA. They are particularly useful because of the way in which they only cut DNA at certain sequence-specific sites. In the bacteria themselves, some of the DNA bases in these same sites on the bacterial chromosome are modified by the addition of methyl ($-CH_3$) groups, which protects them from degradation by the cell's own endonucleases.

Several hundred restriction enzymes have now been identified and isolated. They are each named after the bacterial strain from which they were derived – for example:

1 EcoR1 is from *Escherichia coli*, strain RY13
2 HindIII is from *Haemophilus influenzae*, strain Rd
3 BamH1 is from *Bacillus amyloliquefaciens*, strain H.

Most restriction enzymes recognise and cut DNA within specific nucleotide sequences, often 4 or 6 base-pairs long, called **restriction sites**. These restriction sites are usually in the form of palindromes; that is, they comprise nucleotide sequences that are symmetrical about an axis, and read the same from opposite directions in the two strands of DNA. EcoR1, for instance, makes double-stranded cuts in the way shown in Figure 18.1. It produces a cut that is staggered, giving **sticky** or **cohesive ends**.

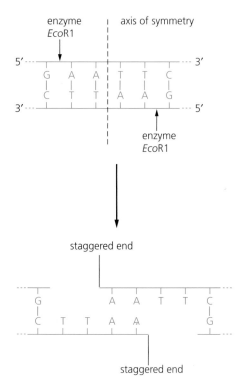

Figure 18.1

Other restriction enzymes, such as HaeIII, cut the DNA strands at positions directly opposite one another, giving **flush** or **blunt ends**.

Palindromes are found in the DNA of all species. They are located at particular sites that are randomly distributed along the chromosomes. As far as we know, they have no particular biological function. Their occurrence, however, is extremely useful to the genetic engineer, because they are the sites that are cut, or cleaved, by restriction enzymes. If samples of DNA from two different species are treated with the same restriction enzyme, say EcoR1 again, then they will both be cleaved at the *same* palindromic restriction sites and will have the *same* sticky ends to their fragments. When the two kinds of DNA fragments are mixed together, they will anneal (by complementary base-pairing) and some new kinds of molecules will be produced. The rejoined ends can then be sealed by special joining enzymes, called **ligases**. These joining enzymes can also be used for sealing blunt-ended fragments.

This capability to fragment DNA into discrete pieces and then to join different fragments together forms the basis of **recombinant DNA** technology (Figure 18.2). It means that we can manipulate DNA and move genes around from one species to another, in order to engineer new kinds of organisms that will perform certain desired functions.

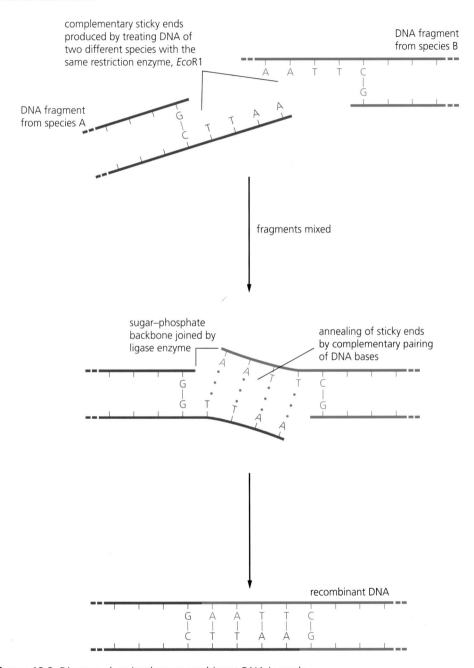

Figure 18.2 Diagram showing how recombinant DNA is made

Synthesis of gene products in bacteria

Plasmids are small circular molecules of double-stranded DNA found in many kinds of bacteria, in addition to their main chromosomes. They replicate themselves independently of the main chromosome. In general, they are not vital to the cell and can be used by the genetic engineer as **vectors** to carry fragments of DNA, or genes, from other species. Plasmids can be taken out of bacterial cells, modified in various ways as recombinant DNA molecules, and put back into bacteria that serve as hosts, to allow them to multiply, or express their genes.

A generalised scheme for making recombinant DNA plasmids in E. coli is illustrated in Figure 18.3. The sequence of operations begins by obtaining plasmids from a culture of bacterial cells. The bacteria are broken open and the plasmids separated out by centrifugation. These plasmids usually contain single sites for a number of different restriction endonucleases. They also contain one or more genes for antibiotic resistance. The bacterial cells with the plasmid genes for antibiotic resistance will grow on a medium containing the antibiotic; those lacking the resistance genes will die. Cells that carry the plasmid can therefore be selected for by growth in a medium containing the antibiotic.

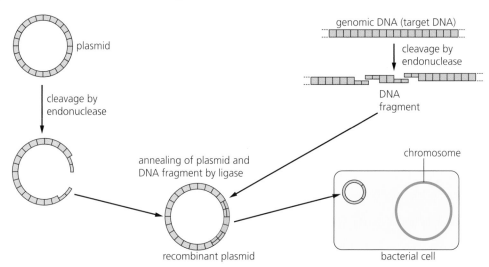

Figure 18.3 Generalised scheme for making recombinant DNA plasmids in the bacterium E. coli

The plasmids are opened up at one of their restriction sites, by cleaving with the appropriate restriction enzyme. They are then mixed with the desired DNA fragments, which have been produced using the same restriction endonuclease. Because both plasmid and target DNA have identical sticky ends, they form recombinant DNA molecules, some of which contain a single fragment of the target DNA (Figure 18.3). After their single-stranded sticky ends have been annealed, the gaps in the sugar–phosphate backbone of the DNA are sealed by DNA-ligase. These recombinant plasmids are then mixed with host bacterial cells, some of which take in plasmids through their cell membrane. This transformation is aided by treatment with $CaCl_2$, which makes the host cells 'competent' for transformation. A number of cultures are then started from single cells on media containing antibiotic, so that those that contain recombinant plasmids can be sorted from those that carry non-recombinant plasmids. We will not go into the details of how this sorting out is done.

There are many valuable gene products that it is desirable to obtain in large quantity. Some of them are of commercial value, such as enzymes used in industry, while others have important medical applications; for example, insulin, interferon and somatostatin. These proteins can now be synthesised inside genetically-engineered bacterial cells and then extracted by various means. The human growth hormone somatostatin is an important example. It is a small protein of only 14 amino acids. The gene was synthesised in the laboratory and then inserted into a plasmid of *E. coli* in such a way that it could be 'switched on' and made to produce the human growth hormone inside the bacterial cells (Figure 18.4).

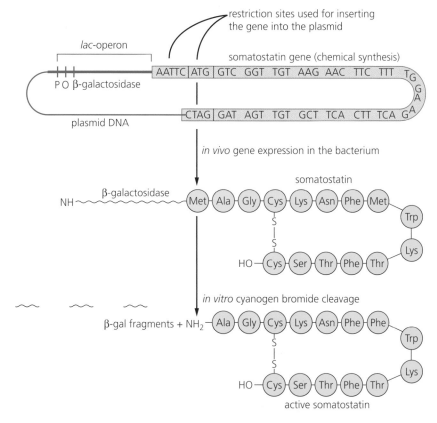

Figure 18.4 Synthesis of the human growth hormone somatostatin using a synthetic gene inserted into a genetically engineered plasmid of *E. coli*. The plasmid also contains part of the lactose operon of *E. coli* that will 'switch on' the gene when lactose is fed to the cell culture. After synthesis, the cells are broken open to release the proteins and a simple chemical treatment then splits off the somatostatin from the β-galactosidase that is made with it. The somatostatin gene is represented as the *genomic sequence* – this is the sequence in the non-transcribed strand of the DNA, which has the identical codons to those of the mRNA except that **T** is present in place of **U**. (Based on K. Itakura *et al.*, 1977, *Science* **198**, 1056–1063)

Transformation in eukaryotes

Eukaryotes are much more complex organisms than prokaryotes and it is correspondingly more difficult to modify their genetic material by recombinant DNA technology. The objectives are simple enough – insert genes into animals and plants that will make them more efficient producers or give them better heritable resistance against pests and diseases. It is even possible to modify plants and animals so that they produce useful products, such as biodegradable plastics and pharmaceuticals. Ultimately it may become possible to modify the genomes of humans so that serious genetic disorders can be treated or cured by **gene therapy**.

In engineering new forms of plants and animals there are three stages that must be accomplished.

1 The desired gene must be cloned and then multiplied up in a bacterial culture to give an abundant supply. Methods of obtaining genes are discussed on page 248.
2 Placing the gene into the recipient nucleus. In the case of animals, a fertilised egg may be manipulated *in vitro* and then re-implanted into a foster mother. Alternatively, sperm may be transformed and used for *in vitro* fertilisation or artificial insemination. The egg cells of plants cannot be so easily manipulated as they have a rigid cellulose cell wall. One way around this problem is to work with **protoplasts**. These are cells that have had their cell walls removed by enzymes (Figure 18.5). In some species, single protoplasts can be grown in culture and regenerated into whole plants (for example, tobacco).
3 Integrating the new genes into the recipient's own DNA, ideally at a particular site where they will be stably inserted, and will function to make their protein products. Transformed individuals may not have the new gene in *all* of their cells; then they are said to be **mosaic**. If some of the germ line cells (eggs, sperm or pollen) are transformed then the new genes will be transmitted to progeny during the normal course of reproduction. From this progeny individuals can be selected that do carry the inserted DNA in all their cells.

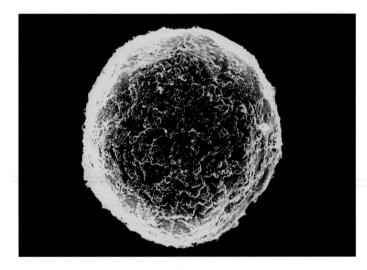

Figure 18.5 An example of a plant protoplast as seen under the scanning electron microscope (×3000). This one is from a moss (*Physcomitrella patens*). The protoplast has been released by treatment of the 'leafy' tissue of the moss with an enzyme solution that strips off the cellulose walls from the cells. It has been cultured for four hours in a growth medium and is regenerating a cell wall

Genetic engineering of animals

One method of animal transformation involves simply mixing DNA fragments with eggs or sperm to get passive uptake of the genes. Sperm is used for *in vitro* fertilisation or for artificial insemination. Eggs are re-implanted into the mother, or implanted into a foster mother. These methods are simple, but also inefficient since only a small proportion of eggs or sperm will actually take up the DNA.

The main method of transformation therefore involves micro-injecting genes directly into eggs. Usually, the DNA is micro-injected into the male nucleus of a fertilised egg cell just before the male and female nuclei fuse together (Figure 18.6). In some animals, such as fish, this is not possible and the genes may then be injected into the cytoplasm of the egg cell, or the nucleus of the zygote.

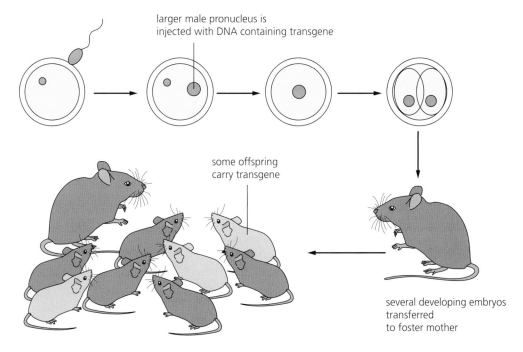

larger male pronucleus is
injected with DNA containing transgene

some offspring
carry transgene

several developing embryos
transferred
to foster mother

Figure 18.6 Micro-injection is a commonly used technique for introducing DNA into animals. Shortly after fertilisation has occurred, and before the male and female nuclei have fused, DNA is injected into the larger male pronucleus using a fine glass needle. Developing embryos are then transferred into a foster mother. The injected DNA becomes integrated into the genome in only some of the eggs, and is lost from others, meaning that only some of the offspring born will carry copies of the transgene

In other animals, especially birds, certain viruses are used as vectors for the transport of **transgenes**. They naturally insert their DNA into the host chromosomes, taking the new genes with them (Figure 18.7, overleaf). Once inserted the viral DNA is replicated along with the chromosomal DNA. The hazard with this approach is that viruses themselves may do harm to the recipient host cells.

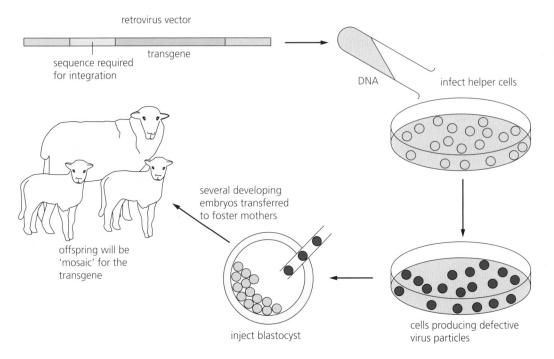

Figure 18.7 Another method for introducing DNA into animals is to use disabled viral vectors. The vector is genetically engineered so that it contains only the sequences required for the integration of viral DNA into the host chromosome, and the desired transgene. The disabled vector DNA is unable to form virus particles. 'Helper' cells are therefore used to assemble the DNA into virus particles, which are then injected into developing embryos (called blastocysts). The DNA is integrated into cells of the blastocyst with high efficiency. The developing embryos are then transferred to foster mothers. The offspring resulting from this procedure may be mosaic, in that only some of their cells carry transgenes. However, some will produce gametes containing the transgene, and some of *their* offspring will therefore carry transgenes in *all* their cells

An example of a successful attempt at genetically engineering a new strain of an animal is that of the 'Mighty Mouse'. A gene known as **GH**, which produces a growth hormone in the rat, was cloned and joined up with a promotor region of another gene that is switched on by the presence of certain metal ions in the diet. The new gene was then micro-injected into fertilised mouse eggs, which were subsequently re-implanted into foster mother mice. Transformed mice were found among the progeny, with the new genes integrated into their chromosomal DNA. Rat growth hormone was actively produced in various tissues of the body and some of the mice grew to twice their normal size. Some were also able to transmit their new genes to their offspring when mated with normal strains of mice.

Several fish species have also been genetically engineered with growth hormone genes, so that they may grow larger and mature more quickly. Especially in salmon, some spectacular increases in growth have resulted, with genetically engineered fish being an average of 4–11 times the weight of non-genetically engineered individuals of the same age.

Cloning

Although **cloning** is itself not a technique of genetic engineering, it is now being used to produce domestic animals that produce valuable proteins. This includes, for example, the cloning of genetically engineered sheep that produce the blood clotting factor VIII used for the treatment of some kinds of haemophilia.

Cloning has been going on since life began. In its simplest form, cloning is the production of cells with the same genetic information; for example, micro-organisms that reproduce by mitotic division are clones of one another, as are plants produced from cuttings or runners. In molecular biology the term is also used to refer to the production of identical DNA molecules, either in bacteria or, as discussed in Chapter 19, using enzymes in a test tube.

With the exception of identical twins and triplets, animals cannot usually be cloned naturally, although it has long been possible to clone some animals, such as frogs. This involves removing the nucleus of an egg cell to produce an **enucleated** egg cell. Either this egg is then fused with a cell from the animal to be cloned, or a nucleus from such a cell is injected directly into the egg cell with a fine glass needle. Depending on the species of animal, the 'hybrid' egg cell may then need to be transplanted into a real or surrogate mother. Such techniques have been used to produce livestock of a particularly desirable genotype, and also in the conservation of rare species.

Until recently, it was only possible to clone mammals by taking a nucleus from embryonic cells. In 1997, though, Dolly the sheep was born – the first mammal to be cloned using a nucleus from an adult. Dolly was followed by Cumulina, the first of three generations of mice produced by injecting adult cell nuclei into enucleated eggs. The birth of Dolly caused a sensation. It was described in *Science* magazine as the '*technological breakthrough of the year*' and speculation abounded about whether we will soon have the technology to clone humans, something already banned in Britain.

The possibility of cloning cells with nuclei from adults is, though, potentially important for another quite sensational, but more morally acceptable reason. The advance brings us nearer to the possibility of being able to clone replacement organs for transplants. If such technology became possible it would be very beneficial. Currently, it is difficult to do transplants because our bodies treat donated organs as foreign bodies. Transplanted organs are attacked by the immune system, which therefore needs to be suppressed using drugs, leaving the body vulnerable to opportunistic infections. If replacement organs could be cloned from the cells of patients needing transplants they would match the original organs, except of course that they would be non-faulty. This could allow transplant surgery without the risks of rejection. There would be no need for immunosuppressant drugs, and less risk of infection for transplant patients.

Genetic engineering of plants

Many species of plants are attacked by a bacterium called *Agrobacterium tumefaciens*. This is a soil-borne pathogen that enters wounds and causes the formation of tumours called **crown gall tumours** (Figure 18.8). In addition to its main chromosome, *A. tumefaciens* carries a large plasmid called the **Ti plasmid**. When infection takes place, part of this plasmid (called the tumour-inducing DNA or **T-DNA**) becomes integrated into the host cell chromosomes. This T-DNA, in fact, is the only part of the pathogen that is transferred into the host. Once integrated into the host chromosome the T-DNA transforms the infected cells and exploits them to produce its own gene products – including tumour-specific compounds such as opines and nopalines. These tumour-specific products are then secreted from the tumour cells and used as a source of nutrition by the *Agrobacterium* cells living around the tumours.

Figure 18.8 Crown gall tumour on a plant stem, produced by *Agrobacterium tumifaciens*

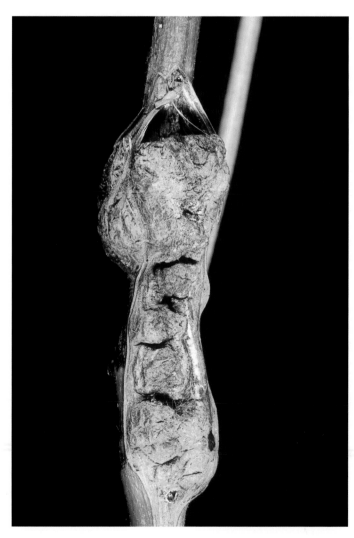

Genetic engineers make use of this natural system of transformation as a means of placing desired genes into crop plant genomes. Another *Agrobacterium* species, *A. rhizogenes*, is also used. This induces a condition known as 'hairy root disease'. The T-DNA of the vector plasmids is modified so that:

1 the tumour-inducing genes are removed
2 the T-DNA retains its capacity to integrate into plant chromosomes
3 the T-DNA carries new genes to modify the plant.

In order that the T-DNA of the engineered plasmids is stably integrated into the chromosomes in all of the cells, and can be inherited as a permanent part of the genome, whole plants are regenerated from transformed tissues.

Although both species of *Agrobacterium* are able to infect a wide range of dicotyledonous plants, cereals and grasses are monocotyledons and it is extremely difficult to transform them using these bacteria as vectors. Alternative methods of transformation were therefore developed for the genetic engineering of these crops. One method, called **electroporation**, uses shorts bursts of electricity to cause protoplasts to take up DNA directly. Another widely used procedure, called microprojectile bombardment, involves the use of a 'DNA gun' to shoot DNA-coated tungsten or gold particles into plant cells.

Many genetically engineered crops have been modified to be resistant to particular herbicides, especially glyphosate. Such herbicides are called broad-spectrum herbicides because they control a wide range of weeds. They are less environmentally damaging than many other available herbicides and, because they inhibit an enzyme that does not occur in animals, they are safer too. They are, however, also deleterious to many crop plants. Genetic engineering of some of these crops involved introducing a gene isolated from glyphosate resistant bacteria. Glyphosate resistant cotton is now commercially grown in America, while other crops are being developed and tested in many parts of the world. Other crops have been genetically engineered to produce their own insecticide, for enhanced nutritional quality, and for improved tolerance to stress.

In addition to modifying plants for agricultural purposes, genetically engineered plants are being developed to produce products that would not normally be produced in plants, including industrial enzymes, plastics and pharmaceuticals. Plants are being developed to produce some of the most expensive drugs available, such as granulocyte-macrophage colony-stimulating factor, which is used for the treatment of impaired human immune systems. Other plants have been genetically engineered to produce vaccines against a variety of human and animal diseases, including cholera and hepatitis B. Much effort is going into the development of edible plants that produce vaccines, which can be distributed to developing countries by the provision of plants or seeds.

Genetically modified crops

The first **genetically modified** plants to be ready for commercial planting were genetically modified food crops, which are now widely grown in countries such as China, America and Canada. Most common are crops modified to be herbicide tolerant, or to produce their own insecticide. In Europe, genetically modified crops undergo extensive safety testing before their cultivation is allowed outside of especially constructed glasshouses and growth rooms. Even then, field trials may be necessary before large scale cultivation is allowed. In the UK, as in other parts of Europe, even the field trials have caused concern among the public and media, many of whom doubt the safety of genetically modified crops.

To have an informed opinion about genetically modified crops, one must consider what is known about their safety. The possible risks include the following. Genetically modified crops may:

1 escape agriculture and become weeds
2 adversely influence the survival of other organisms in the environment
3 give rise to foods that are harmful to human or animal health.

The first two of these are true for any organisms we intentionally or accidentally introduce into the environment. This includes conventionally bred crops and exotic garden plants, which have caused many weed problems and much environmental damage world-wide. Other organisms can also cause damage; for example, escaped pet cats in Australasia, and accidentally introduced diseases such as Dutch elm disease. Although genetically modified crops are less likely to cause problems than many other organisms we introduce into the environment, they undergo considerably more safety testing than most.

It is possible that a few crops, such as forages and sugarbeet, could pass genes to wild relatives. This might be undesirable if, for example, the genes made weeds tolerant of herbicides. However, there are several methods of preventing such gene flow, including preventing flowering by using sterile crops, or by harvesting before flowering. Most crops are unable to cross-breed with wild plants, and many cannot themselves survive outside of agriculture.

Some crops are modified to be tolerant of herbicides used for weed control. It has been argued that the cultivation of such crops might increase the use of chemical sprays. There are, however, cases where the introduction of herbicide resistant crops could reduce the use of harmful herbicides – by reducing the number of applications necessary, or by allowing the use of herbicides that break down rapidly, instead of spraying with more persistent and damaging herbicides such as atrazine and 2,4-D.

It is very unlikely that transgenic DNA will affect humans who eat genetically modified crops. DNA is not toxic and is degraded when eaten. We all eat miles of DNA every day, including genes that code for toxins; for example, in (non-genetically modified) potatoes. In addition, we routinely eat the entire genome of many organisms such as apples, oranges, bananas, oysters and even cattle (rare steak!), not to mention the numerous microbes that can be found everywhere. Transgenic DNA is not fundamentally different to the DNA in other foods.

Some transgenic proteins could be toxic or cause allergies, just as other proteins could. This will be especially true for genetically modified crops that produce pharmaceuticals or enzymes for industry. However, in such cases it is likely that the crops will be grown on a small scale, probably in glasshouses, and will be carefully monitored to ensure that they do not enter into animal or human food chains.

There is the potential to use genetically modified crops to great advantage. Rice, for example, provides more than half the daily food for about a third of the world's population. Some genetically modified rice tested in China, Korea and Chile has 35% bigger yields than conventional rice, and extracts as much as 30% more carbon dioxide from the atmosphere, which could help reduce global warming. In addition, genetic engineering can help to make rice more nutritious by introducing genes for the production of vitamin A. This vitamin is normally lacking in rice and, as a result, some people for whom rice is the staple diet suffer from a dietary deficiency that causes blindness. Many thousands of people could be prevented from going blind in the future if they switch from growing normal rice to growing such genetically modified rice.

There is no reason to suppose that genetically modified food plants are any more likely to cause damage to the environment than other crops, and in some cases they may be beneficial. In addition, genetically modified foods and other crops can be used for substantial benefit to human health, not only by allowing farmers to produce larger yields of more nutritious crops, but also through the production of medicines.

Human gene therapy

There are many human diseases that are caused as a result of gene mutation. In the future, it may be possible to treat, or even cure, such diseases using human gene therapy. The idea is to take cells from a patient and transform them with a normal, non-mutant, copy of the gene causing the disease. The cells are then re-introduced into the patient, where they produce the required protein.

The first use of gene therapy in humans was in 1990 by French Anderson and Michael Blaese. They treated a four year old girl called Ashanti DiSilva, who was suffering from the disease **adenosine deaminase deficient severe combined immunodeficiency disease** (ADA SCID). Sufferers of ADA SCID lack the enzyme adenosine deaminase and toxic levels of its substrate (deoxyadenosine) build up. This kills T-lymphocytes, which are white blood cells involved in the immune system. Minor infections can be serious or even fatal and weekly injections of the enzyme are required if the sufferer is not to be confined to living in a germ-free bubble. At the time of the gene therapy, Ashanti had been permanently confined to quarantine to reduce the chances of her becoming infected with the common childhood diseases.

ADA SCID was a good candidate for the first attempt at gene therapy for several reasons. The gene involved was one of the first disease genes to be cloned and characterised. White blood cells are easy to remove and can be transformed with a working copy of the ADA gene, before being returned to the patient by transfusion. Then, even small amounts of the ADA enzyme the gene produces are beneficial (and overproduction of ADA is apparently non-toxic). Within a year Ashanti's bloodstream had normal amounts of disease-fighting white blood cells. The treatment has to be repeated every few months, because of the short life-span of white blood cells, so the little girl was not cured. She was, however, able to go to a normal school and take part in many normal childhood activities, which she had not been able to do before the treatment began.

Since this initial treatment, much research has been directed at developing and improving the methods of gene therapy, both for ADA SCID and other diseases. In particular, much effort is going into developing gene therapy for cystic fibrosis, some forms of diabetes, various inherited blood disorders, and certain types of cancer, including cancers of the brain, breast and lungs.

Obtaining genes

Eukaryotic genomes are extremely large and contain many millions of nucleotide base-pairs. How can we find and isolate a single gene from so much DNA?

One way is not to try, but to create our own copy of the gene from scratch. In the case of a small protein, such as somatostatin or insulin, the amino acid sequence of the polypeptide is known and this tells us the genetic code of the gene concerned. The gene sequence is then made by chemical synthesis. Automated machines are available that can be programmed to assemble DNA nucleotides to order, and can be used for the manufacture of small genes. Once a few copies have been made they can then be cloned in the usual way. In time it may be possible to synthesise any gene we require by this means.

Another method of obtaining individual genes is by identifying and isolating their mRNAs. In certain cells of an organism, only one gene, or a few genes, may actually be working and producing mRNA. This is the case in the reticulocytes of the bone marrow, which carry mainly the mRNA for haemoglobin. In such cases, the mRNA can be extracted and then used as a template on which to make a complementary DNA (cDNA) by reverse transcription. This is done with the aid of an enzyme called **reverse transcriptase,** which will produce a single-stranded DNA from mRNA. The cDNA is then made double-stranded with the enzyme DNA polymerase. A cDNA copy of the gene is therefore obtained indirectly by reverse transcription from its own messenger RNA.

There are also ways of sorting out a single desired mRNA from a whole mixture of mRNA transcripts, and it is now possible to obtain cDNA copies of almost any gene, even when its mRNA is not abundant within the cells. In eukaryotes, these messengers will have undergone a 'processing stage' before leaving the nucleus and certain sequences in them, which correspond to the non-coding introns of the genomic DNA in the chromosomes (Chapter 12, page 151), will have been spliced out. Furthermore, the mRNA transcripts do not contain the sequences at the beginning of the gene that are necessary for transcription, because these sequences are not themselves transcribed into mRNA (that is, sites for ribosome binding and the attachment of RNA polymerase). The cDNA copy of a eukaryotic gene is therefore different from that of its genomic sequence.

If the genomic sequence itself is required, it can be isolated from a mass of DNA fragments by using a radioactive cDNA or mRNA **probe**, which will hybridise to the corresponding genomic sequence by complementary base-pairing.

Summary

◆ The discovery of restriction enzymes in the 1970s opened up a completely new field of genetics.

◆ Restriction endonucleases are isolated from bacteria, which use them in defence against viruses that invade their cells. These enzymes cause double-stranded cuts in DNA and when used in conjunction with ligases, which are joining enzymes, they enable us to make completely novel forms of recombinant DNA.

◆ Genes can be removed from one species and inserted into the DNA of another one so that new kinds of genetically engineered DNA molecules, and hence new kinds of organisms, can be made.

◆ Bacterial plasmids have proved particularly useful as vectors for cloning pieces of DNA. They can also carry foreign genes, so that their cells may be used as factories for the manufacture of desirable gene products.

◆ Recombinant DNA technology has been used to transform higher plants and animals, and to modify them to produce better crops and more productive farm animals.

◆ Experimental work is also underway to produce transgenic plants and animals that produce useful compounds, such as pharmaceuticals and biodegradable plastics.

◆ The techniques of genetic engineering are also now starting to be used in gene therapy, and in the future we may be able to use them to correct or treat certain genetic defects in humans.

Further reading

1   *DNA Learning Centre*: A variety of instructional materials and information for both students and teachers: **http://vector.cshl.org**
2   Genetic Alliance – Genetic Resources:
    http://www.geneticalliance.org/resources.html#GENETICS
3   *The Natural History of Genes*. Classroom and home activities, including how to extract DNA. Includes sections entitled 'students and teens', 'genetics and society' and 'genetic disorders': **http://gslc.genetics.utah.edu**

Questions

1 a What is genetic engineering?

 b Give an account of the genetic engineering of a named example.

2 a Describe the methods used to genetically engineer animals.

 b What are the potential benefits and hazards of genetic engineering?

 c How might genetic engineering be used for human gene therapy?

3 Give an account of genetic engineering in plants. Describe the different methods used to modify plants and give examples of plants that have already been modified. What benefits could genetically modified plants have?

4 The diagram below shows how genetically modified bacteria may be produced for the production of human proteins.

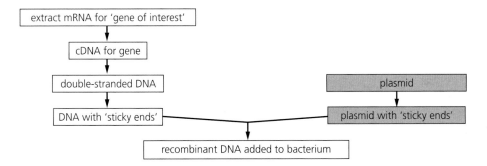

a What is cDNA? Briefly describe the process of producing cDNA from mRNA. State what enzyme is involved in this process.

b What enzyme is involved in producing double-stranded DNA from single-stranded cDNA?

c What is a 'sticky end'? What class of enzymes are used in the production of sticky ends?

d What is a plasmid? Why may the plasmid be described as a 'vector'?

e How are plasmids and human DNA molecules combined and repaired? What enzyme is used to mend the DNA sugar–phosphate backbone?

f In humans, there may be specific DNA that regulates the synthesis of protein produced by the 'gene of interest'. Suggest why human regulator DNA should be replaced by bacterial regulator DNA.

g Why might the bacteria to be transformed be treated with calcium ions and subjected to a temperature shock?

h Describe one way in which bacteria could be screened for the presence of recombinant plasmids.

5 a Explain the use of progeny testing in selective breeding. [8]

 b Distinguish between *selective breeding* and *genetic engineering*. [8]

UCLES A Modular Sciences: Applications of Genetics June 1998 Section B Q1

6 Tobacco plants do not grow well in salty soil. Scientists have used genetic engineering techniques to insert a gene for salt resistance from the bacterium *Escherichia coli* into the DNA of tobacco plants.

  a Describe how scientists could:

    (i) remove the gene for salt resistance from the DNA of the bacterium *Escherichia coli*

    (ii) insert this gene into the DNA of a tobacco plant.   [4]

  b Briefly describe how you could test whether these genetically engineered tobacco plants do grow better than normal tobacco plants in salty soil.   [3]

        AQA (NEAB) AS/A Biology: Continuity of Life (BY02) Module Test March 1999 Q3

7 T-toxin is an insecticidal substance which is produced by a species of bacterium, *Bacillus thuringiensis*. Using the techniques of gene technology, it is possible to isolate the gene responsible for the production of T-toxin from bacteria and to transfer this into tomato plants. An outline of this procedure is shown in the diagram below.

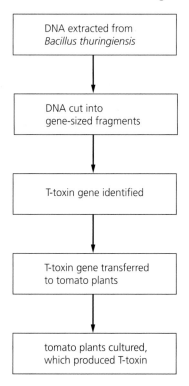

DNA extracted from
Bacillus thuringiensis

DNA cut into
gene-sized fragments

T-toxin gene identified

T-toxin gene transferred
to tomato plants

tomato plants cultured,
which produced T-toxin

  a (i) Name the enzyme which can be used to cut DNA into 'gene-sized fragments'.   [1]

    (ii) Suggest how the gene for T-toxin could be identified.   [2]

  b Suggest how the gene for T-toxin could be inserted into tomato cells.   [2]

  c T-toxin is a protein. Describe the process by which genetically modified tomato plants will synthesise T-toxin.   [4]

  d Suggest *one* possible advantage and *one* possible disadvantage of growing tomato plants into which the gene for T-toxin has been inserted.   [2]

        London AS/A Biology and Human Biology Module Test B/HB1 January 1999 Q6

19

Applications of genetic technology to humans

In addition to being used for the gene therapy, as discussed in Chapter 18, the new techniques of molecular genetics are now being applied to various other aspects of human life, including forensic science, screening for genetic disorders, and medical research. The Human Genome Project, which began in the mid 1980s, was set up to discover the entire genetic code of humans. It is expected that the knowledge from this project will be invaluable for better understanding of inherited disorders, and that we might further develop methods of screening, treatment and gene therapy.

DNA fingerprinting

The technology of DNA fingerprinting was developed by Alec Jeffreys, a British geneticist who was researching the link between DNA variation and exposure to radiation from the accident at the Chernobyl nuclear plant in Ukraine, in 1986. Now the analysis of DNA sequences is often used in forensic science.

By analogy to fingerprinting, where each person's thumb and finger prints are unique, so also is each person's DNA – hence the term **DNA fingerprint**. The only exception to this is in the case of siblings derived from the division of a single egg (identical twins and triplets). Variation is particularly marked in non-coding DNA and DNA not involved in gene regulation; for example, much of the repetitive DNA. This is because mutations in genes and regulatory sequences are mostly selected against, while those in DNA with no function aren't deleterious and can persist.

The uniqueness of each person's genetic fingerprint means that biological samples taken from the scene of a crime can be used to identify, or to rule out, potential suspects. The same technology can be used to determine the relationships between individuals; for example, in a paternity dispute. Very small amounts of DNA can be analysed – from blood and semen stains, for example. The most modern methods can even analyse the tiny amounts of DNA found on hair roots. These methods use a technique called the **polymerase chain reaction** (PCR), which amplifies DNA of the region of interest (Box 19.1). In addition, fingerprinting can be performed on DNA samples that are quite old. It has even been done on the ancient DNA of Egyptian mummies over 2400 years old.

There are various ways of making genetic fingerprints, but all rely on distinguishing between alleles that vary in length rather than sequence, in DNA regions known as **variable number tandem repeats** (VNTRs). A VNTR is a short sequence that is repeated along a length of DNA. It may be a **minisatellite** such as the myoglobin minisatellite, which consists of a 33 base-pair sequence (or unit) repeated four times, or a **microsatellite** with only a short sequence repeated, such as two or three base-pairs. Fingerprinting makes use of such sequences where the exact number of repeats varies between individuals.

VNTR DNA is separated from DNA samples by a PCR reaction, using primers that anneal to sequences either side of the required DNA (Figure 19.1). In addition, restriction enzymes are used to cut up the surrounding DNA but do not make any cuts within the repeat region. Different alleles of the VNTRs can then be seen after being separated from one another by **gel electrophoresis**, a procedure that separates molecules according to their size. During electrophoresis the DNA molecules travel through the pores in the gel, from one end towards the other, under the influence of an electrical current. Small molecules travel quickly while big molecules can only squeeze through larger pores, and travel more slowly, and so less far in a certain time, towards the other end of the gel.

The analysis of DNA fingerprints for criminal cases involves comparing DNA found at the scene of a crime with DNA from the victim and any possible suspects – to see if any of the DNA fingerprints match. The results of fingerprinting are then interpreted using statistics. The frequencies of the different VNTRs in the population have to be determined using sampling techniques. From these can be determined the probability of there being a match between any two people drawn at random from the population. If, for example, the frequency of particular alleles of loci one and two are 1 in 100 and 1 in 50 respectively, then the probability of two people sharing both of these is found from their combined frequencies to be 1 in 5000 (by multiplication). This may not be very convincing evidence, but when more loci are considered the probability of two different people sharing them all can soon become very low. For example, if three more loci were considered in addition, with frequencies of 1 in 10, 1 in 20 and 1 in 55, then the frequency becomes 1 in 55 million. Thus, the chance of anyone having exactly this combination of five alleles is 1 in 55 million (probably only one person in the UK or about five or six people in the USA).

DNA fingerprints have been used as evidence in rape and murder cases in the UK and in many other parts of the world. One of the first such cases occurred in the UK in 1987 when police arrested a young man on suspicion of raping and murdering two teenage girls. The man, who had learning difficulties, first confessed to, but later denied committing both murders. His father suggested that DNA fingerprinting should be done, and this clearly showed that the young man was innocent. It also firmly established that the same person had committed both crimes. The police then arranged for over 5000 local men without alibis to be fingerprinted, but found no match to the DNA sample taken from the scene of the crime. However, someone overheard a man discussing the case in a local pub. The man said he had given a blood sample in the place of a co-worker, Mr. Pitchfork, who had claimed that the police had '*previously treated him unfairly*'. Word got back to the police who arrested Mr. Pitchfork. Although he initially denied having committed the crimes, his DNA fingerprint matched precisely those of the samples from the crimes, and he later confessed. At his trial, the judge said '*The rapes and murders were of a particularly sadistic kind. If it weren't for DNA fingerprinting you might still be at large today and other women would be in danger.*' To that it can be added that the DNA fingerprints probably also prevented an innocent young man from being convicted for a crime he didn't commit.

Box 19.1 The polymerase chain reaction (PCR)

The PCR procedure is a cell-free method of cloning DNA that was invented by the American geneticist Kary Mullis. It is illustrated in Figure 19.1. The DNA sample is first **denatured** (made single-stranded) by heating it to about 95 °C. A heat-stable DNA polymerase enzyme is then used to replicate the DNA strands. However the polymerase can only do this by extending the 3′ end of a **primer** – a short DNA sequence that is bound to the DNA to be copied when the reaction temperature is cooled to about 52 °C (Figure 19.1a). A particular sequence can be replicated by adding primers known to anneal either side of that sequence (Figure 19.1b).

   The process of denaturation, annealing primers, and replicating the single-stranded DNA is known as a 'cycle'. With each cycle the amount of DNA is doubled: after just ten cycles there is more than a thousand times the original amount of DNA. The reaction is carried out in a PCR machine called a **thermocycler**, which can be programmed to carry out as many cycles as a geneticist asks for.

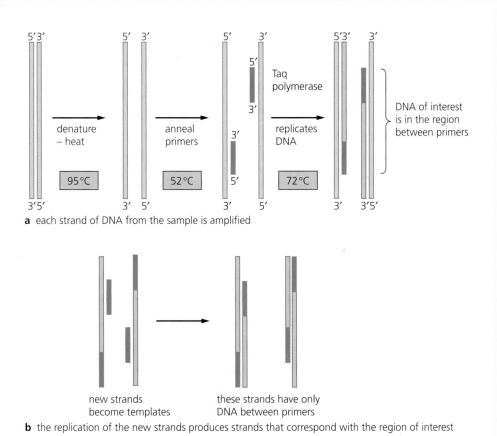

a each strand of DNA from the sample is amplified

b the replication of the new strands produces strands that correspond with the region of interest

Figure 19.1

Screening for genetic disorders

Modern molecular techniques have enhanced the ability to diagnose genetic disorders. In addition to considering clinical symptoms and the results of biochemical tests, doctors can now look at the patient's DNA to see if there are any mutations for particular genetic disorders. In this way they can confirm or rule out a diagnosis made on the evidence of medical symptoms. They can also detect individuals at risk from genetic disorders that occur in adults (called 'late onset diseases'), such as Huntington's disease.

Genetic screening also allows the identification of carriers – people who have one normal and one mutant allele and might pass the mutation on to their children. When both the male and the female in a couple carry a genetic disorder there's a chance of their children being affected. In this case, embryos can be tested at an early stage (prenatal diagnosis).

Cystic fibrosis

Cystic fibrosis is one of the most common inherited disorders, occurring at a frequency of about 1 in 2000–2500 people in Europe, and is more common in people of Northern European descent. The organs of sufferers become clogged with mucus, especially the lungs. This increases the incidence of chronic infections and can lead to organ malfunction. Mucus in the digestive tract can interfere with food absorption and lead to malnutrition. Sufferers often die young and have a poor quality of life.

Although over 500 mutations in the cystic fibrosis gene (*CFTR*) have been identified in different patients, about 70% of cases in Europe are due to a specific three-base deletion mutation in the *CFTR* gene (the mutant allele is called ΔF508). Proteins encoded by the gene are chloride ion pumps, located in cell membranes. As their name indicates, they are responsible for pumping small molecules in and out of cells. Proteins made by ΔF508 have a single amino acid missing, preventing them from being folded during protein maturation. Unfolded proteins cannot take their place in cell membranes and are quickly destroyed by the cell.

The ΔF508 mutation can be detected using a PCR test, which amplifies a 100 base-pair region that includes the site of the mutation (Figure 19.2). This gives a way of detecting the allele in people affected by, or carrying, cystic fibrosis.

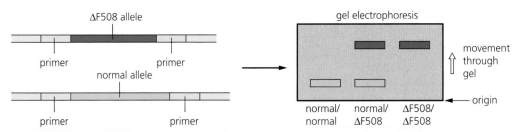

Figure 19.2 Chromosomal DNA is amplified using PCR, with primers that anneal either side of the site of the ΔF508 mutation. The PCR products are separated using gel electrophoresis. As normal alleles are three base-pairs bigger than the mutant allele they migrate more slowly through the gel. The three possible genotypes can therefore be clearly distinguished according to the number and positions of visible bands

The Human Genome Project

The **Human Genome Project** is an internationally co-ordinated programme to determine the entire sequence of the human haploid genome. It grew initially out of a collaboration between scientists at the US Department of Energy (DOE) and the US National Institute of Health (NIH). In 1988, the Human Genome Organisation (HUGO) was set up to co-ordinate research on the Human Genome Project in laboratories worldwide. Two years later a 15 year plan was agreed, to be completed in a series of stages, summarised below. However, rapid technological advances have accelerated research and the expected completion date is now 2003.

1 A **genetic map** has been constructed giving the order, and distances between, linked genes and molecular markers along the length of each of the human chromosomes. Genetic distances are based on recombination frequencies between genes and other markers, there being little recombination between genes closely linked on a chromosome and much more between genes far apart (Chapter 6).

2 **Physical maps** are being made of each chromosome. These maps show precisely where genes and molecular markers are on the chromosomes.

3 Human DNA is being cut up into chunks and engineered into vectors to make what is called a DNA library. The vectors are used to transform bacteria, which can be cloned indefinitely to create copies of the library (Chapter 18). The DNA chunks are being put into the order in which they occur along the different chromosomes. Once the identities of the chunks are known they will be useful for research, including the isolation and then sequencing of disease genes.

4 The ultimate goal of the Human Genome Project is to sequence all of the haploid human genome, but starting with genes of medical significance. This part of the project has included developing faster methods of DNA sequencing and new computer technology for the storage and analysis of the huge amount of data that is being generated by the project. The completion of a 'draft sequence' of the human genome was announced in June 2000. This 'draft' comprises a scaffold of known DNA sequences across an estimated 90% of the human genome. It is of lower accuracy than the finished sequence will be, and is not continuous across the three billion bases of the human genome. Nevertheless, the data will be invaluable to many researchers who are attempting to understand genes of biomedical importance. Some chromosomes have been almost completely sequenced to high accuracy. This includes the completion of chromosome 21 in May 2000. The chromosome is the smallest human chromosome and causes Down syndrome when trisomic (see page 182).

5 In addition the genomes of other simpler model species are being sequenced. The small genomes of several viruses and bacteria have already been sequenced, as have the three larger genomes of baker's yeast (*Saccharomyces cerevisiae*), the roundworm (*Caenorhabditis elegans*), and the fruit fly (*Drosophila melanogaster*). It is expected that by first understanding the simpler genomes it will be easier to work out the complexities of the structure and function of the human genome.

The Human Genome Project is now well under way and all of the targets set for the period to 1998 have been met or exceeded. It is expected that the entire genome will be sequenced by 2003, which is two years earlier than initially expected (and will coincide with the 50th anniversary of Watson and Crick's discovery of the structure of DNA). The Human Genome Project is having, and will continue to have, a great impact on our understanding of genetics, and on the development of medical research and health care. Even when the human genome is sequenced, though, much work will be needed to understand the functioning of the genome, and the proteins it produces. Some of the things that will be made possible as a result of the Human Genome Project include the following.

1 Genes will be found and isolated more efficiently, due to the availability of better genetic and physical maps. This includes genes that cause genetic disorders.
2 Once we know the sequence of normal working human genes, it will be easier to determine which genes are mutated in people with genetic disorders. Knowing the sequence of mutant alleles for genetic disorders will allow the development of genetic tests for them; for example, to identify people who might be at risk from the disorder, or at risk of passing the disorder on.
3 Once the sequences of genes are known the structure and function of the proteins they code for can be worked out. This will help us to better understand how proteins interact, and what goes wrong to make people ill. Drugs can then be developed that can either target particular proteins, or mimic them, thereby adjusting protein function in a way that is beneficial to human health. Eventually, it may even be possible to develop methods of gene therapy, so that working copies of mutated genes can be introduced directly into patients.

Ethical and social issues

Genetic screening technology is developing at a fast pace. It may soon be possible to test individuals for many disorders at once, including some that a person may not develop for years or even decades. Such information could radically alter our reproductive decisions and improve medical care. It does, however, also leave us with unresolved legal, social and ethical issues. For example, what should people know before they have a genetic test? Some late onset diseases, such as Huntington's disease, are not only currently incurable but it is also not always possible to alleviate suffering. Should people be told if they will develop such a disease? Perhaps it will influence their decision about having children. (In the case of Huntington's disease the mutation causing the disease is dominant, and so will eventually be expressed even if only one copy is present.) Perhaps also it could contribute to years of depression before the onset of the disease. Alternatively some people may wish to know so that they can make the most of their health while they have it. Clearly genetic screening should only be carried out in conjunction with **genetic counselling** by someone who explains the meaning of possible outcomes before testing, and who can give support and advice once a result is known.

Genetic testing will be able to identify people who *might* develop a disease, such as people with a genetic predisposition for cancer. This doesn't mean that person will necessarily develop the condition, only that they may be at more risk than average, and perhaps only then if they are exposed to certain other risks, such as a poor diet or lack of exercise. Telling people at risk could help them to alter their life style in a way that might decrease the risk of future illness. It might also make insurance companies or employers less disposed towards insuring or employing such people, if they were allowed to have such information. Generally this would seem to be a bad thing, but one can envisage situations where it might be in the interest of public safety. For example, if an airline pilot is predisposed to heart disease it might be necessary to override the right of the individual to privacy in favour of the right of the public to travel safely. One could argue that there's little difference between this and the similar use of information from eye tests. The right of access to one's personal genetic information will be a matter for consideration from both an ethical and a legal standpoint.

Genetic analysis has shown us that genes affect the risk of developing common diseases and also psychiatric disorders. It is also very likely that our abilities and even personalities are influenced by our genes. Now it is becoming ever more possible to identify, isolate and characterise the genes involved. There is a danger that we will come to see our health, personality and other characteristics as being entirely determined by our genes – the 'deterministic view'. This view is unjustified. Genes interact with both the environment and with other genes. The interactions are complex, making it impossible to predict things like behaviour from a genotype alone. Just because you do or don't have particular genes doesn't mean you will necessarily display certain characteristics, it just makes you more or less likely to do so. The belief that genes are somehow 'super-deterministic' in comparison to the environment is a myth. This fact must be stressed, so that it does not lead to discrimination against people on the basis of their genotype.

During the establishment of the Human Genome Project, concerns such as those discussed above were raised. In order to address these concerns the Ethical, Legal and Social Implications (ELSI) program was established as part of the Human Genome Project. The program also addresses the issue of education, both of the public and of professionals. Nevertheless, decisions taken about such social, ethical and legal issues should not be left to geneticists alone, but should be taken after debates involving the whole of society.

Summary

◆ The applications of modern genetics technology are already diverse. Basic techniques, such as PCR and gel electrophoresis, can be applied in various useful contexts, including DNA fingerprinting and screening for genetic disorders.

◆ The Human Genome Project's major goal of sequencing an entire human genome is likely to be fulfilled in the not too distant future. Along the way much more will be discovered about human genes, especially those causing genetic disorders or otherwise influencing our health.

◆ Even when the complete genome sequence is known a lot of research will be needed to fully understand how the genome works, and the purposes and functions of the proteins it produces.

◆ Even then, we must not forget that genes interact with both one another and with the environment, making it impossible to predict many human characteristics from genotypes alone.

Further reading

1 F. S. Collins (1992) Cystic Fibrosis: molecular biology and therapeutic implications. *Science* **245**: 774–779.
2 Cystic fibrosis: **http://www.kumc.edu/gec/support/cystic_f.html**
3 *Genetics Education Centre:* covers the Human Genome Project, careers in genetics, and offers a wide variety of links to other resources, including lesson plans: **http://www.kumc.edu/gec/**
4 Genome Map – Cystic Fibrosis: **http://www.ncbi.nlm.nih.gov/cgi-bin/SCIENCE96/nph-gene?CFTR**
5 A. J. Jeffreys (1987) Highly variable minisatellites and DNA fingerprints. *Transactions of the Biochemical Society* **15**, 309–316.
6 T. Strachan and A. P. Read (1996) *Human molecular genetics.* BIOS Scientific Publishers.
7 Peter Sudbery (1998) *Human molecular genetics.* Addison Wesley Longman Limited.
8 *Understanding Gene Testing:* Produced by the National Cancer Institute. Includes basic information on 'what are genes' and 'how do genes work': **html://www.accessexcellence.org/AE/AEPC/NIH/index.html**
9 T. Wilkie (1993) *Perilous knowledge: The Human Genome Project and its implications.* Faber and Faber.

Questions

1 **a** Explain the theoretical basis of genetic fingerprinting and suggest uses for the process. [6]

  b Outline how genetic fingerprinting is carried out. [12]

            UCLES A Modular Sciences: Applications of Genetics June 1997 Section B Q2

2 **a** Briefly describe the symptoms and inheritance of cystic fibrosis.

  b Explain how genetic screening can be used to test for cystic fibrosis, and for carriers of the condition.

3 The human gene for alpha-1-antitrypsin (AAT) was introduced into fertilised eggs of sheep and the eggs implanted into surrogate mothers. Some surrogates produced transgenic female animals which secreted AAT in their milk. The sequence of events is shown below.

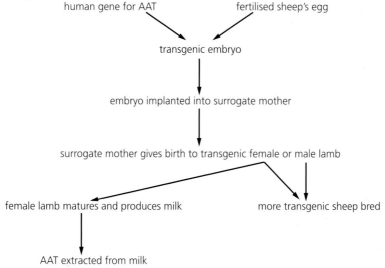

a **(i)** Explain what is meant by a *transgenic animal*. [2]

  (ii) Describe how a transgenic embryo might successfully be implanted into a surrogate mother. [3]

In some cases the injected DNA may not be incorporated into the DNA of the fertilised egg and in others the transgenic female may not produce much AAT in her milk.

b Suggest why the transgenic females produce different quantities of AAT. [2]

Complex human proteins, such as AAT, cannot be produced by genetically engineered bacteria.

c Suggest why bacteria cannot be genetically engineered to produce complex human proteins such as AAT. [1]

A possible use of AAT is the treatment of cystic fibrosis. AAT inhibits an enzyme, elastase, which helps to remove dead or damaged tissue in the body. In people suffering from cystic fibrosis, frequent lung infections lead to the presence of large numbers of white blood cells, which release extra elastase, causing a potentially fatal breakdown of lung structure.

d (i) Describe the inheritance of cystic fibrosis. [3]
   (ii) Explain why sufferers of cystic fibrosis develop frequent lung infections. [2]
In a small clinical trial in 1996–7, patients with cystic fibrosis inhaled AAT into their lungs.
e Suggest how this treatment might help someone with cystic fibrosis. [2]

OCR (Cambridge Modular) A Sciences: Applications of Genetics March 1999
Section A Q4

4 a Describe how genetic screening is carried out. [10]
b Discuss the need for genetic counselling. [6]

OCR (Cambridge Modular) A Sciences: Applications of Genetics March 1999
Section B Q1

5 Thalassaemia is an inherited condition controlled by a single gene with two alleles, the allele for thalassaemia being recessive. It is a disorder which affects the functioning of red blood cells, causing anaemia. This condition was common in Cyprus 25 years ago, but since then the incidence has decreased significantly. This decrease resulted from a programme of genetic screening and counselling.
a Genetic screening involves testing individuals in the population for the presence of the thalassaemia allele. Which genotype would it be important to identify by this process? Explain your answer. [2]
b Explain how genetic counselling might have led to a reduction of the incidence of the disease. [3]
c Suggest why it is unlikely that the allele for thalassaemia will be eliminated from the population. [2]

NEAB A Biology Paper 2 (BY09) June 1996 Q3

6 Describe how the Human Genome Project will contribute to our understanding of genetics, and the possible applications of the knowledge that will be gained from it. Discuss the ethical, social and legal issues that are having to be considered in connection with the Human Genome Project.

Glossary

active site The part of a protein that must be maintained in a particular form for the enzyme to be functional; for example, the part of an enzyme that binds to the substrate.

adaptation The structural or functional features that have evolved in an organism and which enable it to cope better with its environment.

adaptive radiation The evolution of a group of organisms along several different lines, involving adaptation to a variety of environments.

additive gene action Individual genes make an independent additive contribution to an individual's phenotype. Each gene raises or lowers the value of the phenotype on a linear scale of measurement.

allele A particular form of a gene, occupying the same locus (place) on a chromosome as alternative alleles of the same gene.

allele frequency (*or* gene frequency) The relative proportions of the alleles of a gene present in a population.

allopatric Of populations or of species that inhabit separate geographic regions (in contrast to sympatric).

allopolyploid Polyploid with multiple sets of chromosomes from more than one species.

amino acid The basic building block of proteins (polypeptides).

amniocentesis A technique for determining the genotype of an embryo or fetus *in utero* (that is, in the mother's uterus before birth).

anaphase A stage of mitosis or meiosis when the daughter chromosomes are pulled from the equatorial plate to the poles of the cell. Anaphase is after metaphase and before telophase.

aneuploidy Change in chromosome number involving only part of a chromosome set.

antibiotic A substance produced by a micro-organism that is able to inhibit the growth of other microbes, or to destroy them.

antibody A protein (immunoglobulin) that is usually found in serum and whose presence can be demonstrated by its reaction with a specific antigen.

anticodon The triplet of nucleotides in a transfer RNA molecule that associates, by complementary base-pairing, with a specific triplet (codon) in the messenger RNA during protein synthesis.

asexual reproduction Reproduction not involving the formation and union of gametes from different mating types or sexes. Individuals formed by asexual reproduction have the same genetic constitution as the 'parent'; for example, budding in yeast.

assortative mating Non-random mating between partners chosen for some phenotypic similarity.

autopolyploid Polyploid with more than two sets of chromosomes from a single species.

autosomes The chromosomes of a complement other than those involved in sex determination.

auxotroph A mutant strain of an organism that is unable to synthesise a given organic molecule, and can only grow when the required substance is supplied in the food (in contrast to a prototroph).

backcross A testcross in which an individual is crossed with one of its parents.

balanced polymorphism A stable genetic polymorphism that is maintained by natural selection.

basic number (x) The number of different chromosomes in a haploid set.

bivalent Two homologous chromosomes that are paired together during the first division of meiosis.

blending inheritance An early theory of heredity (now discredited) in which it was thought that the characters of an individual were formed as a result of the blending of essences from its parents.

carcinogen An agent capable of inducing cancer.

carrier   An individual with a mutant allele that is not expressed in the phenotype because of the presence of a dominant allele.

cell cycle (*or* mitotic cycle)   The life cycle of an individual cell.

centromere (*or* primary constriction)   A region of a chromosome that becomes associated with the spindle fibres during mitosis and meiosis.

character   Any observable phenotypic feature of a developing or fully-developed individual.

Chargaff's rules   The number of purine bases in DNA molecules equals the number of pyrimidine bases (that is, they occur in a 1 : 1 ratio). In particular, the number of adenine bases equals the number of thymine bases and the number of guanine bases equals the number of cytosine bases.

chiasma (*plural* chiasmata)   The cross-shaped arrangement of the chromatids that is formed at their point of exchange during crossing over. Chiasmata are visible from diplotene until the beginning of anaphase I.

chi-squared test   A test used to determine the goodness of fit of data to the predictions of a hypothesis.

chromatid   Each of the longitudinal sub-units of a duplicated chromosome that become visible during mitosis and meiosis.

chromatin (*or* nucleoprotein)   The stainable material in the nuclei of cells, which is composed of DNA and proteins.

chromomere   One of many bead-like structures visible along a chromosome during prophase of meiosis and mitosis.

chromosome   A thread-like structure found within the nuclei of cells and containing a linear sequence of genes. Eukaryotes have several chromosomes composed of a complex of DNA and protein, whereas prokaryotes have a single chromosome of naked DNA.

chromosome mutation   Abnormality in the number or structure of chromosomes.

chromosome theory of heredity   The theory that chromosomes carry the genetic information and that their behaviour during meiosis provides the physical basis for the segregation and independent assortment of genes.

clone   A population of cells or organisms derived by mitosis from a single cell or common ancestor. Also refers to identical DNA molecules derived by replication of a single progenitor molecule. Due to the asexual nature of clone production, clones of a particular common ancestor all have the same genotype.

cloning   The production of clones.

c-mitosis   The condition where mitosis is blocked at metaphase by the use of the spindle-inhibiting drug colchicine. The lack of spindle fibres results in chromosomes being spread throughout the cytoplasm, instead of being bunched together on the metaphase plate, making c-mitosis ideal for the study of metaphase chromosomes.

co-dominance   A condition in which both alleles of a gene are expressed in a heterozygote, which then exhibits the relevant characteristics of both parents.

codon   A triplet of nucleotide bases in the DNA that codes for a specific amino acid or for polypeptide chain termination during protein synthesis.

complementary base-pairing   The pairing of bases that allows two strands of DNA to form a double helix. Cytosine of one strand pairs with adenine of the other, while thymine in one pairs with adenine of the other.

complete (or simple) **dominance**   If an allele is completely dominant the phenotypes of homozygous and heterozygous individuals are indistinguishable, the dominant allele completely masking the effect of the recessive alleles in heterozygotes.

complete linkage   When there is no exchange of chromatids between maternal and paternal partners in a bivalent, genes show complete linkage; for example, in male *Drosophila*.

continuous variation   Variation not represented by distinct classes; for example, when individuals are of a continuous range of heights, rather than classified as either 'short' or 'tall'. Multiple genes are usually responsible for this type of variation.

crossing over A process of exchange between homologous chromosomes, which may give rise to new combinations of characters. It takes place by the breaking and rejoining of chromatids and leads to the formation of chiasmata.

cross-resistance Resistance that develops in response to one substance but also gives protection against others.

crown gall tumour A tumour induced in the plant hosts of *Agrobacterium*.

C-value paradox The paradox that some very similar animals and plants have unexpectedly large differences in the amount of their genomic DNA, which is surprising if one considers the evolutionary relationships between them.

cytokinesis Division of the cytoplasmic component of the cell and the separation of daughter nuclei, to separate cells.

cytoplasm The protoplasmic contents of a cell, excluding the nucleus.

daughter chromosomes Chromosomes derived from the separation of sister chromatids of a single chromosome during anaphase of mitosis.

deficiency *See* deletion.

degeneracy (of the genetic code) Where several codons may specify the same amino acid.

deletion (*or* deficiency) A mutation involving the loss of genetic material. The size of a deletion may vary from a single nucleotide to a chromosome segment.

deme (*or* Mendelian population) A local community of a sexually reproducing species in which there is random mating, so that each member has an equal chance of mating with any other member of the population.

denaturation Loss of native configuration of a macromolecule; for example, due to heating. Denatured proteins unfold their polypeptide chains and so loose their biological activity. The two strands of a denatured DNA molecule become separated from one another.

dihybrid cross Cross between homozygous parents that differ in two pairs of alelles (**AABB** × **aabb**) (likewise, 'trihybrid' and 'polyhybrid').

diploid Cells, phases of the life cycles, and organisms that are characterised by having two sets of chromosomes.

diplotene A stage of prophase of meiosis I when the chromosomes of bivalents separate from one another at their centromeres.

directional selection Directional selection operates in a changing environment. It results in a reduction in variance and progressive shift in the population mean for the character concerned until a state of adaptation is reached. It is the main form of selection practised by humans in the improvement of domesticated plants and animals.

discontinuous variation Variability that can be grouped into distinct classes; for example, 'tall' or 'short', 'round' or 'wrinkled'.

disruptive selection Disruptive selection favours two optimum classes of phenotype at the expense of intermediates. It occurs in natural populations where two distinct habitats, or different kinds of resources, exist, and in plant and animal breeding where selection is practised for extremes of size or form.

DNA Deoxyribonucleic acid – a polynucleotide in which the sugar residue is deoxyribose.

DNA fingerprint A banding pattern that is specific to a particular individual's DNA after that DNA has been cleaved with restriction enzymes and undergone Southern blot analysis.

dominant An allele that masks the expression of another allele of the same gene. Also, the character produced by a dominant gene.

double helix The three dimensional structure of DNA, as worked out by Watson and Crick in 1953.

Down's syndrome The best known of the autosomal aneuploid conditions. One form is due to trisomy (three copies) of chromosome 21. The phenotype is characterised by mental retardation and certain distinctive physical features.

duplication   A mutation that results in doubling up of part of the genetic material the size of which may vary from a single nucleotide to a chromosome segment.

electroporation   A process used for genetic engineering, where cell membranes are made permeable to DNA by the application of an intense electric current.

endemic (disease)   A disease permanently established in moderate or severe form in a defined area.

enucleated cell   A cell that has had its nucleus removed; for example, by micro-dissection.

epistasis   Where one gene hides the expression of another gene.

equator   The centre of the cell spindle where the chromosomes lie during the metaphase stage of cell divisions.

euchromatin   A chromosome region that has normal staining properties and undergoes the normal cycle of coiling.

eukaryote   A cell or organism that contains a true nucleus enclosed within a membrane.

euploidy   Variation in the number of whole sets of chromosomes.

evolution   The gradual change in the genetic composition of a population over a number of generations.

exons   The parts of eukaryotic genes that correspond to the sequences in processed RNA transcripts. The DNA that codes for proteins.

fertilisation   The fusion of two gametes of opposite sex to form a zygote.

flush or **blunt ends**   The ends of a DNA molecule that has been cut by a restriction enzyme that cuts the DNA strands at positions directly opposite one another.

fossil   The relic or trace of a living organism, which has been preserved by natural processes in rocks of the past.

founder effect   Change in gene frequencies that can occur when a new population is founded by a small number of individuals that are not representative of the genetic composition of the population from which they derive.

frameshift mutation   A mutation resulting from the addition or deletion of one or more nucleotides, other than in multiples of three, which causes the gene to be misread.

gamete (*or* sex cell)   A mature reproductive cell that is capable of fusing with a similar cell of opposite sex to give a zygote.

gametic number (*n*)   The number of chromosomes in the gametes of a particular organism.

gel electrophoresis   A technique used to separate macromolecules according to their size and electrical charge, as they travel through a gel under the influence of an electric current.

gene   A unit of heredity. A gene occupies a specific site in a chromosome and comprises a segment of DNA double helix about 1000 base pairs long, which codes for an RNA or polypeptide product.

gene flow   The transfer of genes between populations by the movement of gametes, individuals, or groups of individuals, from one population to another.

gene mutations   Sudden heritable changes occurring in genes, giving rise to new alleles.

gene pool   The sum total of the genes within the reproductive cells of all of the members of a deme (Mendelian population).

gene therapy   The treatment of a disease by the genetic modification of the patient's cells.

genetic code   The way in which the genetic information is encoded in the DNA.

genetic counselling   Advice and counselling given about genetic disorders, particularly with respect to screening for, and diagnosis of, genetic disorders.

genetic death   The elimination of segregating individuals that are homozygous for deleterious recessives, or which carry deleterious dominant alleles.

genetic drift   The random fluctuation in gene frequencies due to sampling error.

genetic engineering   The production of recombinant DNA molecules – by the insertion of DNA into a virus, a bacterial plasmid, or other vector system – and incorporation of the recombinant DNA into a host organism.

genetic equilibrium (*or* Hardy–Weinberg equilibrium)   When gene and genotype frequencies in a population are in accordance with the Hardy–Weinberg law.

gene frequency   *See* allele frequency.

genetic load   The extent to which the burden of deleterious alleles causes a population to depart from its optimum fitness.

genetic map   A map showing the relative positions of genes according to linkage studies.

genetically modified   An organism that has had its genome altered by the use of genetic engineering.

genetics   The science if heredity. Geneticists study the transmission, the structure and the action of the material in the cell that is responsible for heredity.

genome   The genes in the basic set of chromosomes.

genotype   Genetic constitution of an individual.

genotype frequency   The relative proportions of the various genotypes, from the alleles of one gene, present in a population.

haploid   Cells or organisms, or phases in the life cycle of organisms, with a single chromosome set.

Hardy–Weinberg law (or **principle**)   A principle that states that in a large, randomly mating population there is a fixed relationship between gene and genotype frequencies and – in the absence of selection, migration and mutation – these frequencies remain constant from generation to generation.

hemizygous   Pertaining to genes present only once in the genotype.

heredity   The process that brings about the similarity between parents and their offspring.

heterochromatin   Chromosome regions that are densely stained.

heterogametic sex   The sex that produces two kinds of gametes with respect to the sex chromosomes.

heterozygote (*adjective* heterozygous)   An individual carrying unlike alleles of a gene (**Aa**).

histones   Basic proteins found in nucleosomes, the fundamental units of structure of chromatin.

homogametic sex   The sex that produces only one kind of gamete with respect to the sex chromosomes.

homologous chromosomes   Chromosomes that are identical in their shape, size and the content and distribution of their genes (although different alleles may be present at the loci).

homology (of genes, DNA or chromosomes)   Sequence similarity that suggests a common evolutionary origin.

homozygote (*adjective* homozygous)   An individual with identical alleles for a particular gene (**AA** or **aa**).

Human Genome Project   An international research effort to determine the sequence of the human genome.

hybrid   An individual resulting from a cross between two genetically unlike parents.

incomplete dominance (semi-dominance, partial dominance)   A condition of a heterozygote in which the phenotype is intermediate between the two homozygous parental forms.

infectious drug resistance   The term used to describe the way in which resistance genes can be transmitted by plasmids during conjugation.

interchange   If two chromosomes exchange segments during translocation this is then sometimes called an interchange.

intergeneric hybrid   A hybrid between two species that are not in the same genus.

interphase   The period of the cell cycle between nuclear divisions.

introns   Non-coding sequences of DNA that are transcribed into RNA.

inversion   The reversal of the gene order, which may result when two breaks occur in the same chromosome.

karyotype   The chromosome complement of an individual defined by the number, the form and the size of the chromosomes at metaphase of mitosis.

Klinefelter's syndrome   A syndrome in humans resulting from the presence of an extra X chromosome in the male (XXY, XXXY).

law of independent segregation   (*or* Mendel's second law) Two or more unlike pairs of alleles segregate independently of each other as a result of meiosis, provided the genes concerned are unlinked.

law of segregation   (*or* Mendel's first law) Contrasting forms of a character are controlled by pairs of unlike alleles that separate in equal numbers into different gametes as a result of meiosis.

leptotene   The first stage of meiotic prophase, immediately preceding synapsis. Chromosomes appear as fine thread-like structures, although they are double because DNA replication has already occurred.

lethal allele   A mutant allele that causes the death of individuals. If the allele is recessive then death occurs for individuals homozygous for the allele.

ligase   A joining enzyme that closes single-stranded breaks in DNA.

light repair system   The system of enzymes that repairs the damage done by non-ionising radiation such as UV light; for example, by the removal of thymine dimers.

linkage   The association of certain genes in their inheritance.

locus   The site in the chromosome where a gene is located.

malignant tumours   Tumours that invade and destroy adjacent tissues, and spread around the body.

mean   The arithmetic average; the sum of a sample of measurements divided by the number of measurements in the sample.

meiosis   The reduction division of the nucleus in which the zygotic number of chromosomes is reduced to the gametic number.

Mendelian population   *See* deme.

messenger RNA (mRNA)   A single-stranded RNA molecule that is formed by transcription and which carries the information encoded in the gene to the sites of protein synthesis on the ribosome.

metaphase   The second stage of mitosis or meiosis during which the chromosomes align at the equator of the cell (metaphase plate).

metastasis   The spread of cancer cells from one part of the body to another.

methylation   The modification of a protein or DNA or RNA base by the addition of a methyl group ($-CH_3$).

micro-evolution   Evolution that occurs quickly enough to be observed within the lifetime of a person.

microsatellite (*or* variable number tandem repeat)   A type of repetitive DNA that contains short repeat sequences. Microsatellites are interspersed throughout the genome and are used as a marker for DNA fingerprinting.

microtubules   Long hollow cylindrical filaments in the cytoplasm that form the mitotic spindle. Made from a protein called tubulin.

migration   *See* gene flow.

minisatellite   A variable number tandem repeat where the repeat unit is larger than for a microsatellite; for example, the myoglobin minisatellite, which consists of a 33 base-pair sequence (or unit) repeated four times.

mis-sense mutation   A base-pair substitution in a gene, which results in one amino acid being changed for another one at a particular place in a polypeptide.

mitosis   Division of the nucleus into two daughter nuclei that are genetically identical to one another and to their parent nucleus.

mitotic cycle   *See* cell cycle.

monohybrid   Progeny of a cross between homozygous parents that differ in one pair of alleles (**AA** × **aa**).

mosaic   An organism that has cells of different genotypes.

multiple alleles   The alleles of a gene that has more than two.

multiple drug resistance (in bacteria) Resistance to several drugs due to the presence of plasmids with several different resistance genes in them.

multiple factors (*or* multiple genes, *or* polygenes) Multiple genes (three or more) with individual small cumulative effects, which control characters showing continuous variation.

multiple genes   *See* multiple factors.

mutagenesis   Mutation caused by a mutagen.

mutagen   An agent (radiation or chemical substance) that is capable of increasing the mutation rate.

mutant   An organism, or a cell, carrying a mutation.

mutation   An inherited genetic change caused by an alteration in the DNA sequence of an allele.

mutation frequency   The frequency with which a mutation is found in a sample of cells or individuals.

mutation rate   The number of mutations occurring in a given gene per unit of time.

natural selection   A process that determines gene frequencies in populations through unequal rates of reproduction of different genotypes.

neo-Darwinism   The re-examination of Darwin's theory in the light of discoveries about the physical basis of inheritance (genes and chromosomes).

non-coding DNA   DNA that is not transcribed into RNA, which is translated into protein. Includes sequences involved in the control of transcription and other sequences that apparently have no function.

non-disjunction   A failure of chromosome separation in which a pair of chromosomes, or chromatids, pass to one pole of a cell instead of to opposite poles.

nonsense mutation   Any mutation (substitution, addition or deletion) that changes an amino acid-specifying codon into a chain-terminating codon.

normal curve   A symmetrical, bell-shaped curve representing the normal distribution and characterised by the mean and standard deviation. Many continuously varying biological characters can be represented by a normal distribution. Approximately 68% of observations lie within one standard deviation of the mean, and approximately 95% within two standard deviations of the mean.

nuclear genome   The total DNA content of the haploid nucleus; that is, not including the genomes of cell organelles such as mitochondria and (in plants) chloroplasts.

nuclear sap   The fluid in the nucleus, which is rich in enzymes and other molecules concerned with the activity of genetic material.

nucleolar organiser region (NOR)   *See* nucleolus organiser.

nucleolus (*plural* nucleoli)   Nuclear organelle in which rRNA is made and ribosomes are partially synthesised. A nucleus may contain several nucleoli, which are usually associated with the nucleolus organiser.

nucleolus organiser (*or* nucleolar organiser region (NOR), *or* secondary restriction) A chromosome region containing the genes for ribosomal RNA.

nucleoprotein   *See* chromatin.

nucleosome   A small spherical body that is the basic unit of eukaryotic chromosome structure. It comprises eight histone molecules encircled by two coils of DNA.

nucleotide   A molecule composed of a nitrogen base, a sugar and a phosphate group – the basic building block of nucleic acids.

nucleus   The membrane-bounded organelle of eukaryotes that contains the chromosomes.

operon   A group if co-ordinately expressed and adjacent structural genes, together with their operator and promotor sites.

pachytene   The third stage of the first prophase of meiosis, when chromosome pairing is completed.

pairing (*or* synapsis)   The coming together of homologous chromosomes during prophase I of meiosis.

panmixis   Random mating within a population of sexually reproducing organisms. Each member of the population has an equal chance of mating with any other member of the population.

partial linkage   Genes located on the same chromosome generally show partial linkage due to a certain amount of crossing over between them.

particulate inheritance   The theory that inheritance is controlled by discrete heritable units (genes).

pedigree   A diagram showing the phenotypes and relationships between family members.

phenotype   The appearance and function of an organism as a result of its genotype and its environment.

physical map   A diagram showing the relative positions of physical landmarks in a DNA molecule. Common landmarks include the positions of restriction enzyme cutting sites and particular DNA sequences.

plasmid   A circular DNA molecule that replicates independently of the main chromosome within a bacterial cell.

pleiotropy   The multiple phenotypic effects of a single gene.

polygenes   *See* multiple factors.

polygenic inheritance   Inheritance of characters whose expression is controlled by several genes with individual slight effects upon the phenotype.

polymerase chain reaction   An *in vitro* method that allows the amplification of DNA using enzymes and primers.

polymerisation   The formation of molecules of repeated units by the linking of those units.

polymorphism   The presence of two or more distinct forms of a species (or of a gene) found in the same locality at the same time, and in such frequencies (>1%) that the presence of the rarest form cannot be explained by recurrent mutation.

polynucleotide   A polymer of covalently linked nucleotides.

polypeptide   A chain of linked amino acids.

polyploid   A cell, tissue or organism with three or more complete sets of chromosomes.

polyribosome   A complex of two or more ribosomes associated with a mRNA molecule and actively engaged in the synthesis of polypeptides. Also called a polysome.

population   A local community of a sexually-reproducing species in which the individuals share a common gene pool.

population genetics   The branch of genetics that investigates the factors determining the genetic make-up of groups of interbreeding organisms (populations).

predisposition   Some genes cause a predisposition for particular characteristic, making it more likely that the individual carrying those genes will display that characteristic; for example, cancer. The effects of the genes may be modified or overridden by other environmental factors.

primary constriction   *See* centromere.

primary structure (of proteins)   The specific sequence of amino acids in a polypeptide chain gives a protein molecule its primary structure.

primer A short sequence of nucleotides that possesses a free 3′ terminal hydroxyl group for the extension of a new DNA polymer.

probe A fragment of labelled DNA or RNA used to detect the presence of complementary DNA or RNA sequences.

prophase The early stage of nuclear division (mitosis and meiosis) in which the chromosomes condense and become visible under the light microscope.

protein A molecule composed of one or more polypeptide chains.

protoplasm Term for the substance of a cell, which can be differentiated into the nucleus and cytoplasm.

protoplast A plant cell without its cellulose cell wall.

prototroph A strain of organism that will proliferate on a minimal medium (in contrast to an auxotroph).

pure line Descendants, by self-fertilisation, of a single individual that is homozygous at all of its gene loci.

pure-breeding (true-breeding) **strain** A group of identical individuals that always produces offspring of the same phenotype when intercrossed.

qualitative variation The existence of a range of phenotypes for a particular character that are distinctly different from one another; that is, show discontinuous variation. Such traits are controlled by a few Mendelian alleles.

quantitative variation The existence of a range of phenotypes for a particular character that differ by degrees on a continuous scale rather than by distinct differences. Such traits are usually determined by many genes.

quaternary structure (of proteins) The interaction of two or more polypeptide sub-units to form a complete protein.

race A genetically or geographically distinct subgroup of a species.

reading frame The way in which the genetic code is mRNA is read as triplets by the ribosome.

recessive An allele whose effect is masked by the presence of a dominant allele of the same gene. Also, the character produced by a recessive gene in the homozygous state (**aa**).

reciprocal crosses Two crosses done with identical genotypes. Each genotype is represented by a female in one cross and a male in the other. Reciprocal crosses are done to test whether the sex of parents influences the outcome of the cross.

recombinant An individual with a new set of characters arising by recombination.

recombinant DNA Novel DNA sequences spliced from DNA derived from more than one source.

recombination The process by which new combinations of parental characters arise.

replication The synthesis of DNA.

restriction endonuclease An enzyme that recognises and cuts a specific sequence of DNA. Many of these enzymes are extensively used in genetic engineering.

restriction enzyme A nuclease that makes double-stranded cuts in DNA at specific palindromic sites.

restriction site Specific nucleotide sequence in DNA that is recognised and cut by a restriction enzyme.

reverse transcriptase An enzyme that catalyses the synthesis of a DNA strand from an RNA template.

RNA A single-stranded ribonucleic acid nucleic acid that is similar to DNA but having the base uracil (**U**) in place of thymine (**T**), and ribose sugar instead of deoxyribose.

ribosomal RNA (rRNA) A class of RNA molecules that are present in ribosomes.

ribosome A complex structure composed of RNA and proteins, which catalyses the translation of mRNA.

RNA polymerases A family of large enzymes that catalyse the synthesis of an RNA strand from a DNA template.

R-plasmid A bacterial plasmid that carries drug resistance genes. Commonly used in genetic engineering.

same-sense mutation   A base substitution in a DNA triplet that does not alter the amino acid sequence of a polypeptide.

sampling error   The chance deviation from a theoretical expectation for some character in a population; for example, the expected frequency of a particular allele of a particular gene, seen in a sample of a population.

secondary constriction   *See* nucleolus organiser.

secondary structure (of proteins)   The interaction between amino acids in a polypeptide chain, causing, for example, the polypeptide to coil up into α-helices, or to form β-pleated sheets.

segregation   Separation of pairs of unlike alleles (**A** and **a**) into different cells as a result of meiosis.

semi-conservative replication   The accepted model of DNA replication. Each double stranded molecule of DNA results from one parental strand and one newly synthesised strand.

sex cell   *See* gamete.

sex chromosome   The chromosomes in the complement that carry the sex-determining genes.

sex-linkage   The location of a gene in a sex chromosome.

sickle-cell anaemia   A severe anaemia in humans. It is caused by an autosomal recessive mutation that results in an amino acid replacement in the beta-globin chain of haemoglobin. Heterozygous individuals tend to be more resistant to malaria than people homozygous for the normal (non-mutated) allele.

somatic cell   A 'body cell', which is not destined to become a gamete.

speciation   The process by which new species arise.

species   Groups of interbreeding natural populations that share common morphological characteristics and which are reproductively isolated from other such groups.

spindle   A barrel-shaped structure made of microtubules, which is formed in a cell during division of the nucleus, and which serves to align and move the chromosomes at metaphase and anaphase.

stabilising selection   Selection against the extreme variants for character (that is, those furthest from the mean), leading to a reduction in variance without any change in mean. It is the form of selection that operates in a constant environment to maintain the best adapted genotypes within the population.

standard deviation   A measure of the amount of variation about the mean in a sample of observed measurements. In normally distributed quantities approximately 95% of all values lie within two standard deviations of the mean.

sticky or **cohesive ends**   Single-stranded ends of DNA fragments produced by certain restriction enzymes. Sticky ends are capable of re-annealing with complementary sequences in other such strands. The production of sticky ends is used to join DNA for genetic engineering.

sympatric speciation   Speciation that arises from an isolating mechanism within a gene pool in the same geographic region.

synapsis   *See* pairing.

tautomeric shift   A reversible change in the location of a hydrogen atom in a molecule, altering the molecule from one isomeric form to another. In nucleic acids the shift is typically between a keto group (*keto* form) and a hydroxyl group (*enol* form).

T-DNA   A segment of a Ti plasmid of *Agrobacterium* species that codes for plant hormone production, which causes tumour induction in plants. T-DNA is flanked by a region that allows it to become integrated into its hosts DNA upon infection.

telophase   The final stage of nuclear division (mitosis or meiosis). During telophase the chromosomes unwind and reform an interphase nucleus.

tertiary structure (of proteins)   The folding or coiling of the secondary structure to form a globular molecule; for example, by the formation of disulphide bonds between sulphur-containing amino acids.

testcross A cross in which a heterozygote is crossed with a corresponding recessive homozygote (**Aa** × **aa**).

thermocycler A machine that can be programmed to carry out the polymerase chain reaction (PCR).

thymine dimer A pair of adjacent chemically bonded thymine bases in DNA. The repair processes that replace the thymine dimer often make errors that lead to mutations.

Ti plasmid A plasmid of *Agrobacterium* species that enables the bacterium to infect plant cells and induce tumour production.

transcription The process in which mRNA is synthesised from a DNA template.

transfer RNA (tRNA) Small RNA molecules that carry specific amino acids to the ribosome during protein synthesis.

transgene A gene from one organism that has been modified by recombinant DNA technology and re-introduced into a different organism.

transgenic organism One whose genome has been modified by the introduction of transgenes.

transient polymorphism The gradual replacement of one allele for a gene by another one.

translation The process by which the transcribed information carried in the base sequence of mRNA is used to produce a sequence of amino acids in a polypeptide chain.

translocation The transfer of a segment of one chromosome to another, non-homologous one.

transposable elements (*or* transposons) A genetic unit that can exit from its position in the genome and reinsert itself somewhere else (that is, relocate itself).

transposons *See* transposable elements.

tubulin The protein that forms the microtubules of the mitotic spindle.

Turner's syndrome A sex chromosome abnormality in humans in which there is only a single X chromosome. Affected individuals are phenotypically female and usually have underdeveloped gonads.

universal donor People of blood group O, who can donate blood to people of any other blood group. There are no A or B antigens on the donors red blood cells that could agglutinate antibodies in the recipients serum.

universal recipient People of blood group AB, who can receive blood from people of any other blood group. There blood has no anti-A or anti-B antibodies that could be agglutinated by donor antigens.

variable number tandem repeat *See* microsatellite.

variation Differences between individuals in a population.

vector A vehicle for transferring DNA between individuals; for example, a plasmid chromosome used in genetic engineering.

wild type The prevailing genotype that is found in nature, or in a standard experimental strain of a given organism.

XYY syndrome Males with an extra Y chromosome. They are usually fertile and may be normal in phenotype, although the condition is often associated with some degree of mental handicap.

zygote A cell formed by the fusion of two gametes.

zygotene The second stage of prophase I of meiosis when pairing (synapsis) takes place between homologous chromosomes.

zygotic number ($2n$) The number of chromosomes found in the zygote of an organism, and in the somatic cells derived from it.

Index

transmission of genetic material 6

transposable elements (transposons) 167

tree diagrams 40

triplets 159

Triticum aestivum 179

Triturus cristatus 153

tubulin 12

Turner's syndrome 181

ultraviolet light 13, 168

units of heredity 123, 150

universal donors 105

universal recipients 105

uracil 142

variable number tandem repeats 253

variation 1, 113–21

vectors 238

warfarin resistance 226–8

Watson, James 6, 126

Wilkins, Maurice 126

X-linked mental disorders 86

X-ray crystallography 126–9

XO system 77–8

XX/XY system 78–9

XYY syndrome 182

Y-linked genes 86

zygote 2

zygotene 18, 19, 20, 21

zygotic number 15